Evolutionary Structural Optimization

Springer-Verlag London Ltd.

Y.M. Xie and G.P. Steven

Evolutionary Structural Optimization

With 134 Figures

Springer

Y.M. Xie
Victoria University of Technology
Department of Civil and Building Engineering, Victoria 8001, Australia

G.P. Steven
University of Sydney
Department of Aeronautical Engineering, New South Wales 2006, Australia

Additional material to this book can be downloded from http://extras.springer.com.

ISBN 978-1-4471-1250-1 ISBN 978-1-4471-0985-3 (eBook)
DOI 10.1007/978-1-4471-0985-3

British Library Cataloguing in Publication Data
Xie, Y. M.
Evolutionary structural optimization
1.Structural design
I.Title II.Steven, Grant P.
624.1'771

Library of Congress Cataloging-in-Publication Data
Xie, Y. M.
Evolutionary structural optimization / Y.M. Xie and G.P. Steven : with 144 figures.
p. cm.
Includes bibliographical (p.) references and index.

1. Structural optimization. I. Steven, G. P. (Grant P.)
TA658.8.X54 1997 97-13040
624.1 - - dc21 CIP

Originally published by Springer-Verlag London Limited in 1997
Softcover reprint of the hardcover 1st edition 1997

Typesetting: Camera ready by authors

69/3830-543210

Preface

In recent years the finite element analysis (FEA) has become a widely used tool for engineers of many disciplines. Structural optimization, however, has achieved far less popularity in practice despite the extraordinary progress of the optimization theory and associated algorithms over the last three decades. This situation is caused, at least partly, by the difficulties and complexities of existing structural optimization methods.

This work attempts to bridge the gap between FEA and structural optimization by introducing an entirely different approach to structural optimization. It is based upon the simple concept of slowly removing (or shifting) inefficient material from a structure so that the resulting shape of the structure evolves towards an optimum. We call this *evolutionary structural optimization* (ESO).

The ESO method is presented here for a whole range of structural design problems including statics, dynamics and buckling. Compared to other methods for structural optimization, ESO is overwhelmingly attractive due to its simplicity and effectiveness. Anyone who has a basic knowledge of FEA can easily understand and apply the ESO method.

This book is written for engineers and researchers, both academic and industrial, with an interest in structural design. Their disciplines include civil, mechanical, and aeronautical engineering. The material contained in this book will also be of interest to biologists, medical practitioners, and others who are interested in the shapes of natural load carriers such as bones, trees, and shells.

One chapter of the book describes the software Evolve97 which has been developed by Mr. O.M. Querin in collaboration with the authors. This software can be used by the readers to either try the ideas of ESO themselves or solve their optimization problems. Evolve97 is provided on a computer diskette, which is included with the book.

The bulk of the material presented in this book is very much the result of a joint effort of two research groups, one at The University of Sydney and the other at Victoria University of Technology. The authors would like to thank all those who have contributed to this work, particularly Mr. D.N. Chu, Ms. E.L. Dunn, Dr. B.G. Falzon, Mr. A. Hira, Ms. D. Manickarajah, Mr. J. Mechler, Mr. J.J. Medved, Mr. B.B.C. Mendis, Mr. O.M. Querin and Dr. C. Zhao. Thanks are also due to Mr. Gerard Carè who provided the authors with access to part of the source code of STRAND6 software and gave much assistance. The authors are grateful to Dr. H. Guan who read the whole manuscript and made many valuable suggestions for improvement.

The authors would like to express their thanks to many friends and colleagues in other parts of the world, particularly from Europe, who have given much support to their work in the past four years. Finally the authors wish to acknowledge the financial support from the Australian Research Council which made this work possible.

Mike Xie and Grant Steven
January 1997

Conditions for Program Usage

1. The disk may only be used by the book purchaser on his/her microcomputer.

2. A back-up copy of the disk may be made by the book purchaser.

3. The program may not be transferred onto any other machine without the written permission of the authors, or in the case of their death, by the written permission of their estates.

4. The software is supplied without any warranties of any kind, whether implied or otherwise.

5. All queries regarding function, performance, or adjustments to the programs should be directed to the authors.

Contents

Chapter 1 Introduction

1.1 Background

Structural optimization is a fusion in the areas of engineering, mathematics, science and technology that has the goal of achieving the best performance for a structure, be it a bridge or a space vehicle or a spectacle frame. All aspects of the environment the structure lives in must be accommodated even undesirable ones, crash worthiness of cars being an example.

Because of the mathematical complexities the topic of structural optimization remained to be of more academic interest until last twenty years. Since then there has been a re-focusing on the topic. This is being pulled by the large number of highly competent analysts who are skilled in Finite Element Analysis (FEA) and who now want to use the method beyond the previous 'checking designs out' capability. At the same time two push factors have changed the topic from the previous narrowness to the stage where it is becoming accessible to the practising engineers and scientists. First the availability of high performance computing power at low cost, and secondly the rapid improvements in algorithms used for design optimization where thousands of design variables and constraints may need to be handled. There have been more and more research activities directed towards making these developments available to engineers and scientists in an easy, reliable, inexpensive and understandable form.

Several classes of methods of structural optimization are proving useful in this emerging role. The classical design parameter search methods have become more efficient due to improvements in algorithms used in the search engines (Vanderplaats 1984). The next powerful class of methods are generically grouped together with the title Optimality Criteria (Rozvany 1989). More recently the Homogenization Method, pioneered by Bendsøe (1995) has proven to be successful in generating optimum topologies for continuum structures by treating element densities as design variables.

This book presents a new way of addressing the age-old question about how things get to be the shape they are or, in an alternative form, how the structural performance of an object can be as good as possible. The authors have been researching the technique that they have named Evolutionary Structural Optimization (ESO) since 1992 and have come to the conclusion that the method provides the best way of attempting to answer the question above. At the same time ESO can be shown to be capable of solving many problems of size, shape and topology optimization.

The ESO method is based on the simple concept that by slowly removing inefficient material from a structure, the residual shape evolves in the direction of making the structure better. An important feature of the ESO method is that it is easy to understand and learn while at the same time producing reliable results. What is also of importance to note is that all aspects of structural behaviour can be accommodated within the ESO concept. In this book it is shown that the optimality constraints can be stress based, stiffness/displacement based, frequency based, bucking load based, with single or multiple environments and the ESO concept can be applied to all these problems.

Evolution has been with us for billions of years and evolutionary algorithms for structural optimization have been suggested at various times over the last 15 years in the literature. Like all really useful things in technology and engineering the authors make no claim to having invented the ESO concept; rather they consider that their contribution has been to make it work, easily, effectively and comprehensively for a full range of engineering situations.

The ESO techniques are often easy to implement. A sample software Evolve97 is described in Chapter 9 of this book and provided in the attached computer diskette. It is the authors' experience that once the philosophical concept of structural optimization has been presented to a graduating class of engineering students, there is no subsequent difficulty in that class being able to formulate an optimization problem and successfully use Evolve97 to achieve significant improvement on present practice.

Throughout the text the words optimization and optimum structure have been used extensively. In the context of a structure subject to multiple loads and support conditions the final shape is that which best satisfies the constraints, with the degree of satisfaction not necessarily the same for all the constraints. In colloquial terms the final shape could be said to be the optimum compromise and is sub-optimal for any individual single load case or single constraint.

It will become clear to the reader of this text that this is not a textbook full of formal mathematics that validates why ESO produces the quality of result that it does. Instead the authors have tried to demonstrate the usefulness and effectiveness of ESO in two ways: firstly by benchmarking the ESO process against existing results and secondly by examining structural situations where intuition plays a part in assessing the final results.

1.2 The Engineering Design Process

"*Everything that did not happen by accident or coincidence happened by design.*"

This quotation in 1987 from Jans Bernsen, then the Head of the Danish Design Council, serves to illustrate the point that if an object has a practical purpose then its shape is linked to and governed by that purpose.

The engineering design process has the major components of **analysis** and a **data base** interacting with the **design evaluation module** being driven by the **design goals**. Many forms of analyses can be considered as part of a design process including:

Financial;	Environmental;	Acoustic;	Energy;
Marketing;	Thermal;	Electrical;	Structural;
Chemical;	Psychological.		

In the engineering design creation cycle, the data base plays two roles. Firstly it provides data on the environment that the product has to exist in and on the materials that could be used to produce the object. This is a valid use of the data base. The second use is that it can provide information on every other version of the product that has come before, to help answer some questions like:

- What has worked well with the product?
- What has not worked well with the product?
- What the designer could do to make the product better?
- What the users could do to make the product work better for them?

Where the data base ought to become less prescriptive is in telling the designer what to make the product look like, its shape. It is relatively simple and certainly much less creative to use the shape and form of the previous product that satisfies the design goals as a basis for a new design. Followed logically back to initial emergence of the product we see that the shape could have emerged before there was a full understanding of its function or even all the design rules that had been applied. In other words history gives the shape for the design and not the laws of nature, therefore there often may be no overall objective for the structure.

Using the existing shape of any object to derive subsequent designs may be the most expeditious and safe procedure but it does not appear to be the most efficacious or indeed to produce the shape that best satisfies the design objectives. Failure to properly describe the design environment can often be the largest source of design 'grief'. The ESO process can easily be used to prove that if a different environment is specified then a different structural form is produced.

The initiation of a design should be regarded as a design creation rather than a design prescription and that the analysis should be an integral part of the creation and not something that is done after the design has been prescribed. Such retrospective checking leads to what is called the 'over the wall' time line for the design process. In such a design process all activities take place in a serial manner with each group in the design process communicating only by passing all the data 'over the wall' to the next group and then having no more involvement until it comes back 'over the wall'.

Figure 1.1 is a schematic of the engineering design process where the analysis interacts with the creation of the design in a concurrent manner.

In Figure 1.1 it is seen that the analysis and the data base interact at the time of the design creation. This has two immediate advantages. Firstly because the process is much more interactive, the 'over the wall' activity does not happen, thus speeding up the design process. Secondly the design is more directly derived from the underlying principles of the various analyses and not driven by historical practice. This is not to deny any historical influence on a design, but more to say that the rational aspects need to be the primary driver rather than secondary one.

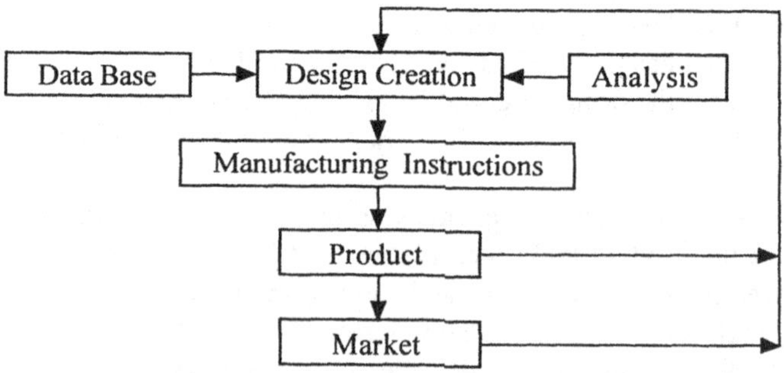

Fig. 1.1. A schematic of engineering design process.

In this book it is shown how the principles and procedures of ESO can be an integral part of this proposed new design cycle whereby an evolving design can be driven by the governing laws and the environment that the product has to live in. In doing so the design process can have no information about the historical shape of the object at all. It is shown in a later chapter that by simply laying down the operational environment for a spanner then applying ESO, a shape is produced which closely resembles what we have come to know as a spanner.

1.3 Illustration: A Simply Supported Beam

Rather than launch into formal mathematics to make a general point, a simple illustration of how structural design affects the usage of material is presented. The ability of a given volume of timber or steel to carry load in the situation of a beam resting on simple supports with a point load at the mid-span, as shown in Figure 1.2, is used as a medium to present part of the message as to why structural optimization is worthy of study.

This is one of the simplest analysis cases presented to undergraduate students whilst studying the topic of the mechanics of solids. In order to maximize the carrying capacity of the beam the engineer can select from a wide range of beam cross sections. This example is to demonstrate how significant the choice of cross section is. It is assumed that the design criterion is that the maximum stress in the beam shall not exceed some material limiting value of σ_{max}. The relationship between the stress σ inside the beam and the bending moment M is given by

$$\sigma = \frac{My}{I}, \tag{1.1}$$

where y is the distance away from the neutral plane (the horizontal mid-plane, in this example) at which the stress is calculated and I is the second moment of area of the cross section. The maximum stress for any section is at the maximum distance away from the neutral plane. The mathematical definition of the second moment of area is

$$I = \int_A y^2 dA, \tag{1.2}$$

from which it is seen that material at higher values of y has a much greater contribution to I than material closer to the origin of y, the neutral plane. For the

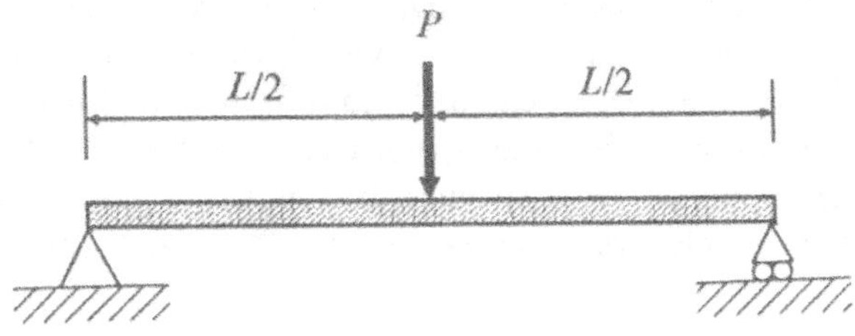

Fig. 1.2. A simply supported beam.

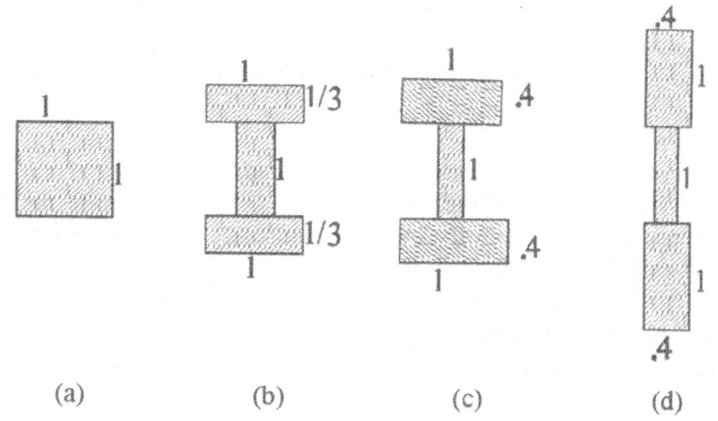

Fig. 1.3. Four possible constant area cross sections.
(all dimensions are times h)

simply supported beam shown in Figure 1.2 the maximum moment is at the mid-span of the beam and is given by

$$M_{max} = \frac{PL}{4}, \tag{1.3}$$

where P is the load at the mid-span and L is the length of the beam. If we combine equations (1.1) and (1.3), an expression for the maximum load capacity of the beam is obtained as

$$P_{max} = \frac{4M_{max}}{L} = \frac{4\sigma_{max}}{L}\left(\frac{I}{y}\right). \tag{1.4}$$

Since the maximum stress and the beam length are fixed the maximum load is obtained by maximizing the value of (I/y). This is determined once the shape of the cross section has been decided. For this illustration the constraint is that the volume of the beam is constant thus the cross sectional area of the beam is also a constant.

If we consider that the basic shape available is a square section of sides h by h then this has the base level of strength against which other possible cross section shapes of the same area can be compared. There are an infinite range of ways the material could be arranged and for simplicity four are presented in Figure 1.3.

Using equation (1.2) above which tells how to calculate the second moment of area for a cross section and equation (1.4), it is not too difficult to evaluate P_{max} for each section and these are presented in the table below.

Table 1.1. Maximum load capacities of the simply supported beam for various cross sections.

Cross Section	P_{max}	$P_{max}/P_{max}(a)$
(a)	$0.6667\sigma_{max}h^3/L$	1
(b)	$1.5852\sigma_{max}h^3/L$	2.378
(c)	$1.8637\sigma_{max}h^3/L$	2.795
(d)	$2.3556\sigma_{max}h^3/L$	3.533

Thus by simple rearrangement of the cross section it is possible to achieve more than 3.5 times the load carrying capacity. Conversely if the load were fixed then we could use much less material to support the same load.

This simple argument is quite convincing as to why we should undertake sufficient analysis in the structural design practice to ensure that material usage is at a minimum. However manufacturing costs have not been included and it could be that to make section (d) the sum of manufacturing cost plus material cost is greater than that for section (a).

Whilst it may seem that the analyst has done an excellent job in deciding on a way of arranging the material that can potentially increase the load carrying capacity of the beam structure, the real story of structural optimization has hardly begun. To realize this, it has to be appreciated that whilst the stress in the outer fibre of the chosen cross section at the load application point must be equal to or less than the limiting value, and for design purposes we set it equal, the rest of the material is at a much lower stress value. This can be appreciated because the stress is linear about the middle neutral plane and the bending moment varies linearly from the mid-span value of *PL*/4 to zero at the ends. With this knowledge it is possible to plot contours of iso-stress relative to σ_{max} on a beam of uniform rectangular cross section. This is shown in Figure 1.4.

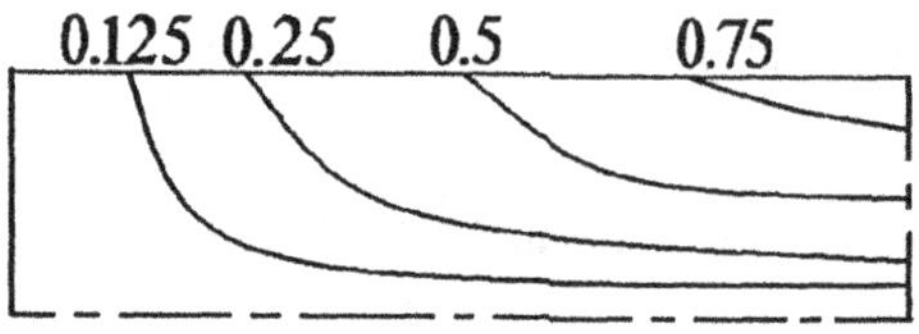

Fig. 1.4. Contours of iso-stress in upper left quadrant of the beam.

It is seen from Figure 1.4 that only a small portion of the beam has a stress greater than 75% of the maximum, indeed this proportion is approximately 3%. In other words the bulk of the material, 97%, is at a stress lower than 75% of the maximum. It is this type of situation that provides the real challenge for structural optimization. The question of how to generate a shape of structure that makes the best use of the material introduces the concept of *fully stressed design* where, ideally, all the material is at the same stress.

The concept of fully stressed design is not new. Indeed there have been contributions to the topic since the 19th century. The most significant early work is that of Michell (1904) on the optimum layout of pin-jointed frames with even stress. This work is well documented in the book by Hemp (1973). For some general background reading on structural optimization the reader is referred to the book by Haftka and Gürdal (1992).

1.4 Wish List for Structural Optimization

If one were to try to write down a wish list for all the aspects that an engineer would be looking for in structural optimization the following would be the main items:

- Size, shape and topology optimization on same problem at different parts of structure.
- Different optimization criteria at different parts of structure, for example, an aircraft having dynamic criteria on the wings, stiffness on undercarriage and fatigue (stress) on fuselage.
- Totally general multiple load environments.
- Totally general multiple support environments.
- Totally general multiple material environments.
- Totally general structural forms in 2D and 3D.
- Optimization across linear statics, dynamics and stability simultaneously.
- Optimization with material and geometric non-linearities.

This is quite a list and is indeed asking that the structural optimization specification to consider the object in the real world rather than a very restricted abstract mathematical one.

1.5 Finite Element Analysis (FEA)

"If there is an opportunity for improving the design, then somebody somewhere is attempting to do so using Finite Elements."

This quotation from Bruce Irons, one of the pioneers of finite element methods, serves to illustrate how important a role FEA is playing in the modern engineering design process. FEA has been around for about 30 years and has now become a mature technique which is taught to nearly all engineering students as a regular part of their curriculum.

FEA is a computational technique which models a structure or a continuum as an assemblage of small parts (elements). Each element is of simple geometry (e.g. triangle or quadrilateral) and therefore is much easier to analyze. The FEA procedure generates many simultaneous algebraic equations, which are solved on a digital computer. FEA originated as a method of stress analysis. Today it has found applications in many other engineering disciplines including heat transfer, fluid flow, electric and magnetic fields. Complicated problems previously intractable by classical analytical methods can now be solved routinely using FEA. Most engineering design offices and manufacturing companies nowadays have FEA computer software in-house or have access to consulting firms that provide FEA services.

The ESO method presented in this book uses FEA as the computational engine. At the end of each FEA the computer assesses the FEA results and then applies simple ESO rules to remove inefficient material. As such, the ESO process is nothing more complicated than running FEA again and again, with the only extra operation of removing some redundant material at the end of each FEA. The subsequent chapters of this book will present ways of identifying where such redundant material is for various design objectives and constraints.

It is beyond the scope of this text to discuss the details of FEA. There are hundreds of FEA textbooks and numerous FEA programs available - a reflection of the popularity of this technique. For those who wish to know more about FEA the following textbooks are recommended: Cook *et al.* (1989), Hughes (1987), and Zienkiewicz and Taylor (1989, 1991).

1.6 References

Bendsøe, M.P. (1995) *Optimization of Structural Topology, Shape, and Material*, Springer, Heidelberg.

Cook, R.D., Malkus, D.S. and Plesha, M.E. (1989) *Concepts and Applications of Finite Element Analysis*, 3rd edn., John Wiley & Sons, New York.

Haftka, R.T. and Gürdal, Z. (1992) *Elements of Structural Optimization*, 3rd edn., Kluwer Academic Publishers, Dordrecht.

Hemp, W.S. (1973) *Optimum Structures*, Clarendon Press, Oxford.

Hughes, T.J.R. (1987) *The Finite Element Method: Linear Static and Dynamic Finite Element Analysis*, Prentice-Hall, Eaglewood Cliffs, NJ.

Michell, A.G.M. (1904) The limits of economy of material in frame-structures. *Phil. Mag.* **8**, 589-597.

Rozvany, G.I.N. (1989) *Structural Design via Optimality Criteria*, Kluwer Academic Publishers, Dordrecht.

Vanderplaats, G.N. (1984) *Numerical Optimization Techniques for Engineering Design: with Application*, McGraw-Hill, New York.

Zienkiewicz, O.C. and Taylor, R.L. (1989) *The Finite Element Method*, 4th edn., Vol. 1: Basic Formulation and Linear Problems, McGraw-Hill, New York.

Zienkiewicz, O.C. and Taylor, R.L. (1991) *The Finite Element Method*, 4th edn., Vol. 2: Solid and Fluid Mechanics, Dynamics and Nonlinearity, McGraw-Hill, New York.

Chapter 2 Basic Evolutionary Structural Optimization

By slowly removing inefficient material from a structure, the shape of the structure evolves towards an optimum. This is the simple concept of evolutionary structural optimization (ESO). Various design constraints such as stiffness, frequency and buckling load may be imposed upon a structure. Depending on the types of design constraints, different rejection criteria for removing material need to be used. For each type of constraints, the corresponding rejection criteria will be discussed in detail in the subsequent chapters. This chapter describes the simplest rejection criterion based on local stress level. Several examples are included to illustrate how the ESO process works.

2.1 Introduction

The process of structural optimization has traditionally followed a parametric path where the shape and topology of a structure are defined by a set of parameters. These parameters are adjusted in order to achieve some objectives, such as minimum volume, but without violating certain constraints. This task has proved difficult and limited success has been achieved using mathematical programming techniques in the last thirty years.

The ESO method offers a new approach to structural optimization, which overcomes most of the problems associated with traditional techniques. ESO is based on the simple concept that by slowing removing inefficient material from a structure, the residual shape of the structure evolves towards an optimum.

A reliable sign of potential structural failure is excessive stress or strain. Inversely a reliable sign of inefficient material use is low stress or strain. Ideally the stress in every part of a structure is near the same safe level. This concept leads to the rejection criterion based on local stress level, where lowly stressed material is assumed to be under-utilized and will be removed subsequently. By gradually removing material with lower stress, the stress level in the new designs become more and more uniform.

2.2 Material Removal Based on Stress Level

First a piece of material which is large enough to cover the area of the final design is divided into a fine mesh of finite elements. Loads and boundary conditions are applied and a stress analysis is carried out using a finite element program. More often than not it is found that part of the material is under-stressed compared to the rest of the structure. Using some criterion for rejection, here called *rejection criterion*, such inefficiently used material may be eliminated.

Since the structure has been divided into many small elements, the removal of material from the structure can be conveniently represented by deleting elements from the finite element model.

The stress level at each point can be measured by some sort of average of all the stress components. For this purpose the von Mises stress has been one of the most frequently used criteria for isotropic materials. For plane stress problems the von Mises stress σ^{vm} is defined as

$$\sigma^{vm} = \sqrt{\sigma_x^2 + \sigma_y^2 - \sigma_x \sigma_y + 3\tau_{xy}^2} , \tag{2.1}$$

where σ_x and σ_y are normal stresses in x and y directions, respectively, and τ_{xy} is the shear stress.

The stress level of each element is determined by comparing the von Mises stress of the element σ_e^{vm} to the maximum von Mises stress of the whole structure σ_{max}^{vm}. At the end of each finite element analysis, all the elements which satisfy the following condition are deleted from the model:

$$\frac{\sigma_e^{vm}}{\sigma_{max}^{vm}} < RR_i , \tag{2.2}$$

where RR_i is the current *rejection ratio* (*RR*).

The cycle of finite element analysis and element removal is repeated using the same value of RR_i until a *steady state* is reached, which means that there are no more elements being deleted at the current iteration. At this stage an *evolutionary rate* (ER) is introduced and added to the rejection ratio, i.e.

$$RR_{i+1} = RR_i + ER\,, \qquad i = 0,1,2,3\ldots\,. \tag{2.3}$$

With this increased rejection ratio the cycle of finite element analysis and element removal takes place again until a new steady state is reached.

Such an evolutionary process continues until a desired optimum is reached, for example, when there is no material in the final structure of which the stress level is less than 25% of the maximum. This may not be the absolute best result, but such an evolutionary optimization procedure offers the designer with many improved solutions, since the shape and topology at each steady state may be chosen as the final design. Ideally the final structure becomes a fully stressed design where the material at each point of the structure is stressed to its full strength. However only in a few special cases can a fully stressed structure be possible.

This evolutionary procedure requires two parameters to be prescribed. The first is the initial rejection ratio RR_0 and the second is the evolutionary rate ER. Typical values of RR_0 = 1% and ER = 1% have been used for many test examples. But for some problems much lower values need to be used. For any new model, after a few trials, it is not difficult to choose suitable values for these parameters. For example, if too much material has been removed from the structure within one iteration or one steady state, it indicates that smaller values should be used for RR_0 or ER.

2.3 Example of Two-Bar Frame

A well-known structural optimization problem is the two-bar frame subjected to a single load F as shown in Figure 2.1. For fixed values of the applied force F and the horizontal length L, the optimal height can be found analytically as $H = 2L$, if the frame is assumed to be pin-jointed.

To obtain such an optimal structural layout using the evolutionary procedure, a model shown in Figure 2.2 is employed where the rectangular design domain is larger than the size $2L \times L$ (L = 10 m). As we have mentioned earlier, the initial model needs to be large enough to cover the area of expected final design.

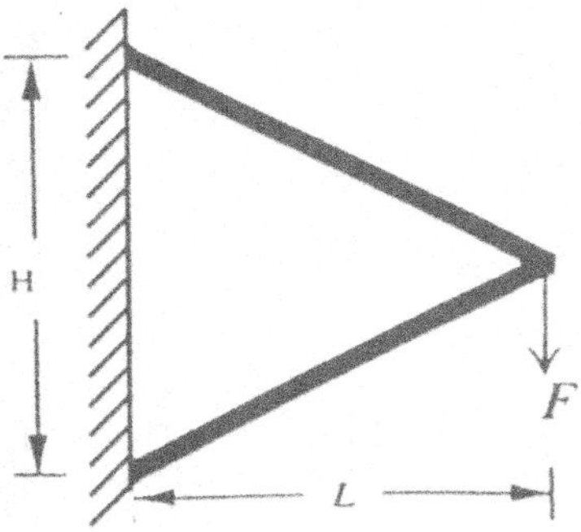

Fig. 2.1. A two-bar frame structure.

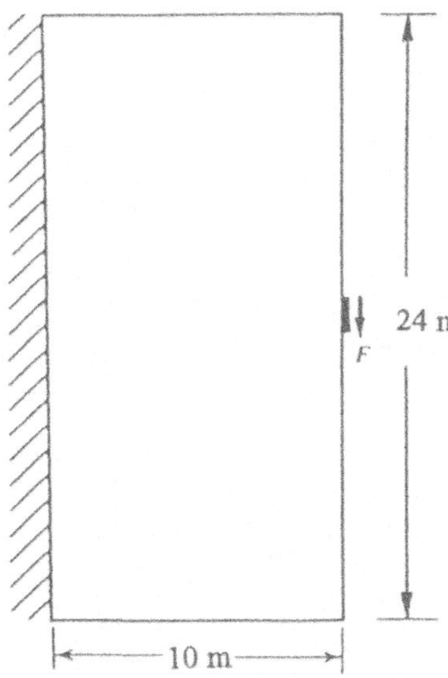

Fig. 2.2. Design domain for the two-bar frame structure.

The design domain in Figure 2.2 is divided into 25 x 60 four node plane stress elements of equal size (0.4m × 0.4m). The thickness of the plate is 1 mm. Young's modulus E = 100 GPa and Poisson's ratio ν = 0.3 are assumed. Shear stress of 1 MPa is applied on the edge of two elements at the centre of the right hand side.

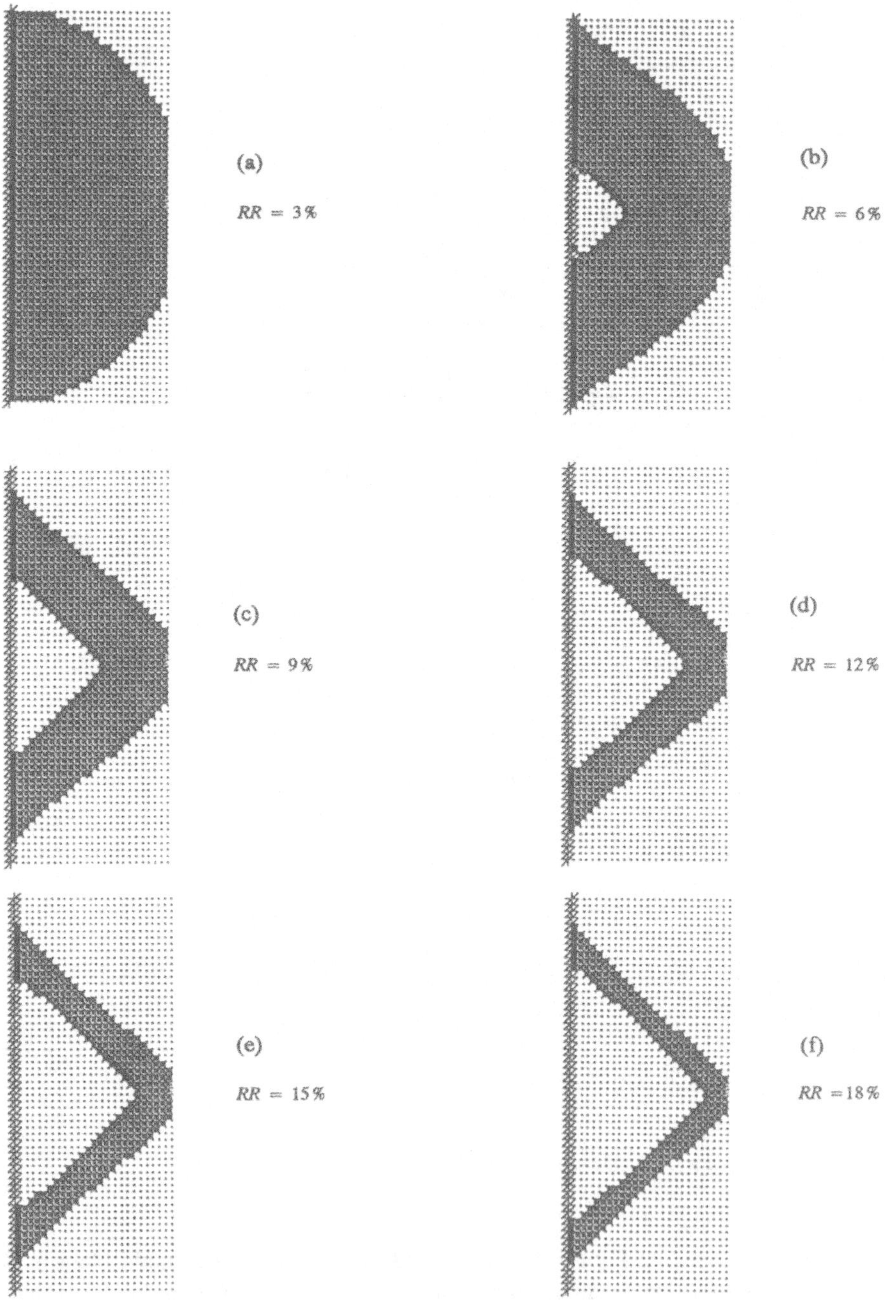

Fig. 2.3(a-f).

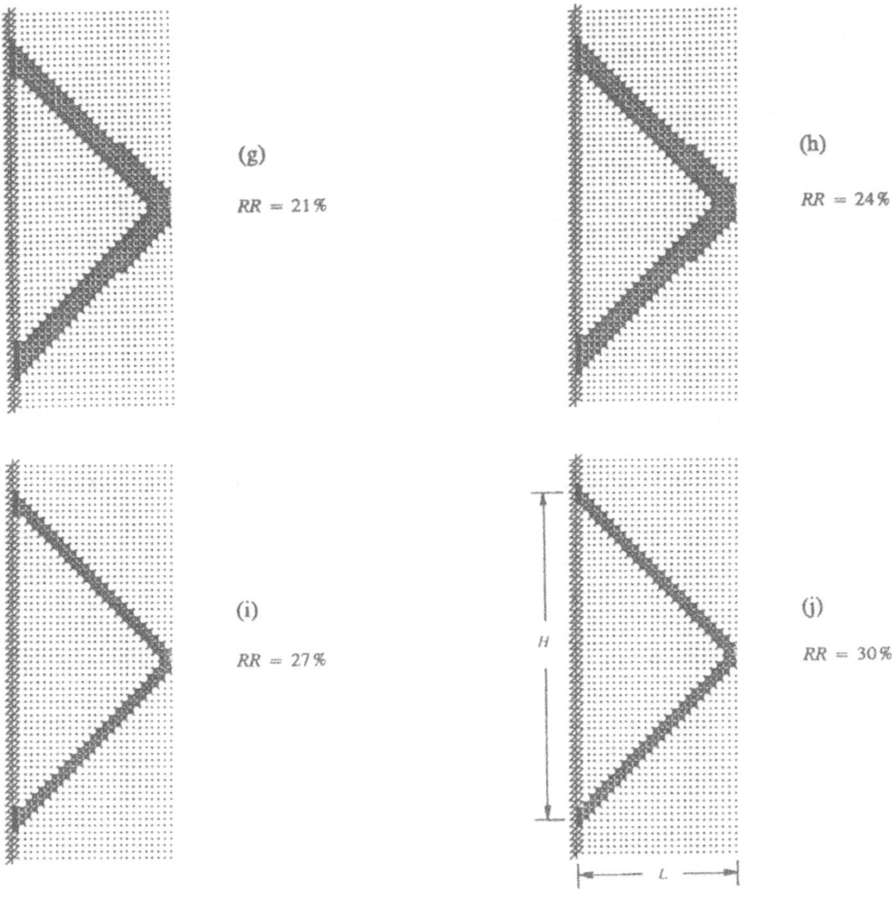

Fig. 2.3. Evolving towards the two-bar frame structure.

We start the evolution with the initial rejection ratio RR_0 = 1%. The evolutionary rate *ER* is also set to 1%. Figures 2.3(a-j) show the different stages of the evolution history. In these figures the black areas represent the remaining elements and the small dots represent the nodes of the initial FEA model. Each of the figures shows a steady state for a certain rejection ratio *RR*. As the rejection ratio increases, more and more relatively inefficient material is removed from the structure. From the early stages of the evolution, it is clear that the structure is to evolve to a two-bar frame. The final result shown in Figure 2.3(j) gives $H = 2L$.

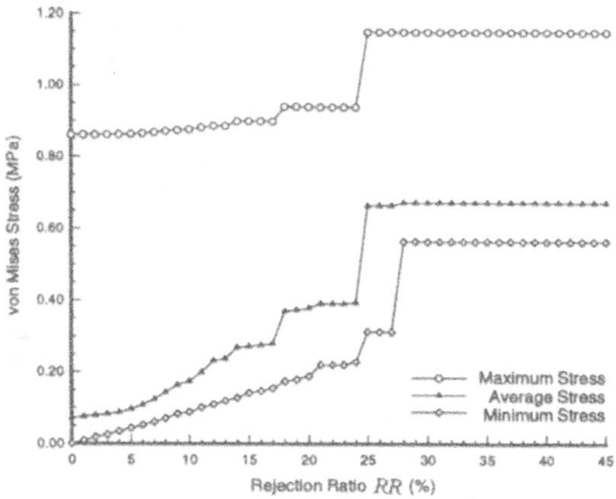

Fig. 2.4. Evolution histories of von Mises stress for the two-bar frame example.

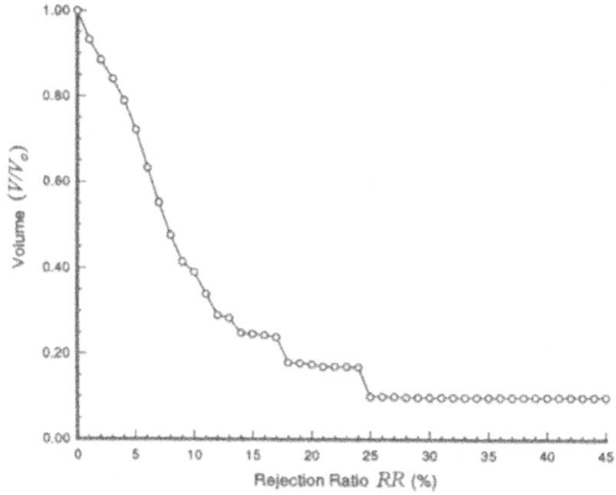

Fig. 2.5. History of volume reduction for the two-bar frame example.

On average eight iterations are needed for the solution to converge to a steady state for each rejection ratio.

The evolution histories of the maximum, minimum and mean values of the von Mises stress are plotted in Figure 2.4. The history of volume reduction is shown in Figure 2.5. Table 2.1 compares the stresses in the initial rectangular model with those in the final design of Figure 2.3(j). It is seen that the huge gap between the minimum and the maximum stresses in the initial model has been significantly reduced. In fact the stress in the final design becomes almost uniform throughout the whole structure except that the stress level of the two elements where the applied shear loading is applied are about twice as high as that of all other elements. When 29% ≤ *RR* ≤ 49%, no material is removed from the model. It is not until *RR* reaches 50% that all the remaining elements are removed from the model, except those two elements with applied shear loading.

In Figure 2.4 it is noticed that when the rejection ratio is raised from 24 to 25%, both the maximum and the mean stresses have sharply increased. This corresponds to a relatively large amount of material being removed from the structure, as can be seen in Figures 2.5 and 2.3.

Comparing the initial model to the final design it is seen that the maximum stress has increased from 0.8603 to 1.1472 MPa (a 33% increase). However, at the same time the total volume of the structure has been reduced by more than 90% of its original volume. The percentage decrease in the volume certainly outweighs the percentage increase in the stress.

Table 2.1. Comparison between the initial rectangular model in Figure 2.2 and the optimized design in Figure 2.3(j)

	σ_{min}^{vm} (MPa)	σ_{max}^{vm} (MPa)	$\sigma_{min}^{vm} / \sigma_{max}^{vm}$	Volume (m^3)
Initial model	0.0003	0.8603	0.0003	0.2400
Optimized design	0.5636	1.1472	0.4913	0.0237

2.4 Examples of Michell Type Structures

The first truss solutions of least weight and a general theory for deriving them were published more than 90 years ago by the distinguished Australian inventor, engineer and mathematician A.G.M. Michell (Michell 1904). One of the most important publications on this topic is the classical book by Hemp (1973).

The truss system shown in Figure 2.6 is one of the Michell structures. It is the least weight truss for the situation of a vertical load acting in the middle of two fixed supports.

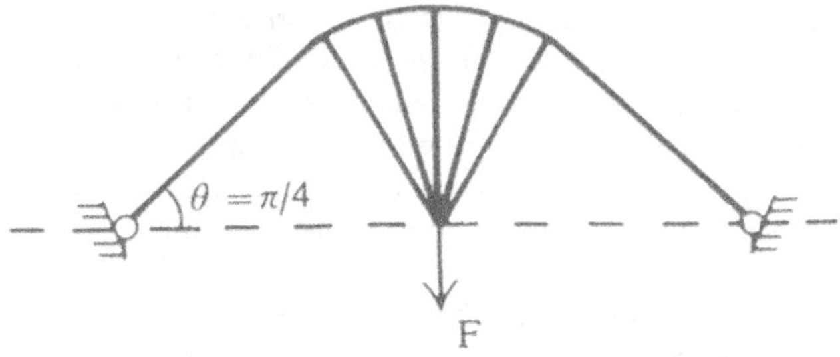

Fig. 2.6. A Michell type structure.

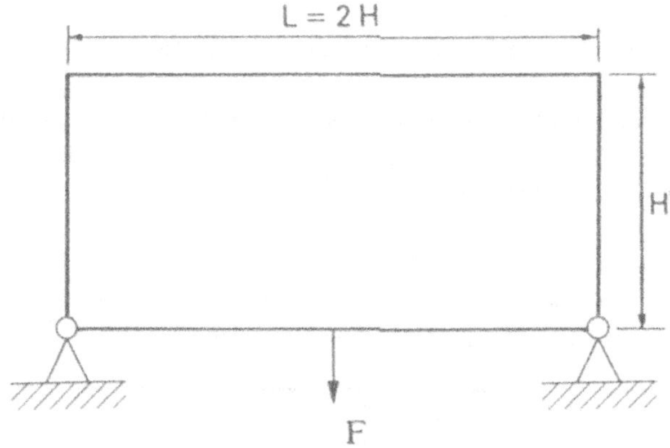

Fig. 2.7. Design domain for a Michell type structure with two fixed supports.

Whether the evolutionary procedure can produce, from a continuum model, a topology similar to Michell's solution poses an interesting challenge.

A design domain of the size $2H \times H$ ($H = 5$ m) as shown in Figure 2.7 is divided into 50×25 four node plane stress elements (in fact, due to symmetry, one needs to analyze half of the model only.). The two corners at the bottom are restrained from translational movements. Young's modulus $E = 100$ GPa and Poisson's ratio $\nu = 0.3$ are assumed. The thickness of the plate is 0.1 m. The vertical load F is equal to 1000 N.

The evolution is started with the initial rejection ratio $RR_0 = 1\%$ and the evolutionary rate $ER = 0.5\%$. Figures 2.8(a-c) show the results of three steady states with the rejection ratio RR being equal to 5, 10 and 15 per cent, respectively. The evolution histories of the maximum, minimum and mean von Mises stresses are plotted in Figure 2.9. The history of volume reduction is shown in Figure 2.10.

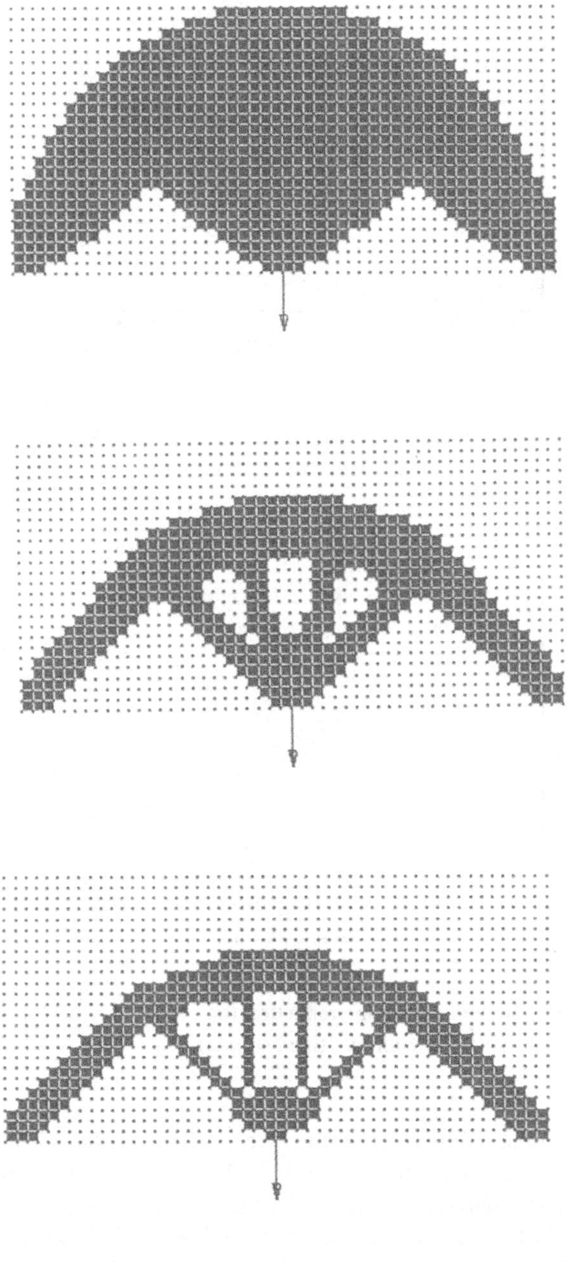

(a) $RR = 5\%$

(b) $RR = 10\%$

(c) $RR = 15\%$

Fig. 2.8. ESO solutions for a Michell type structure with two fixed supports.

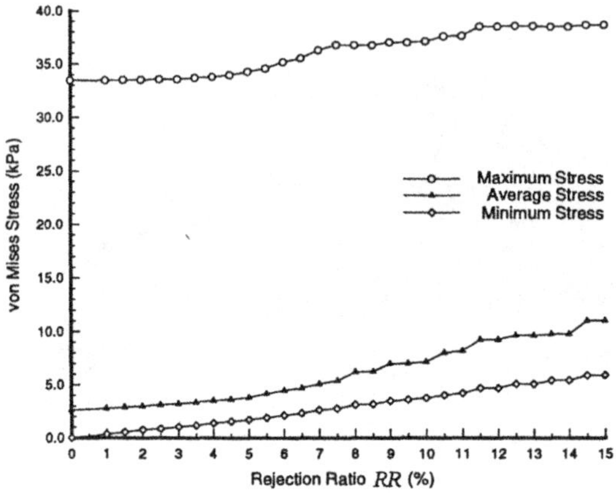

Fig. 2.9. Evolution histories of von Mises stress for a Michell type structure with two fixed supports.

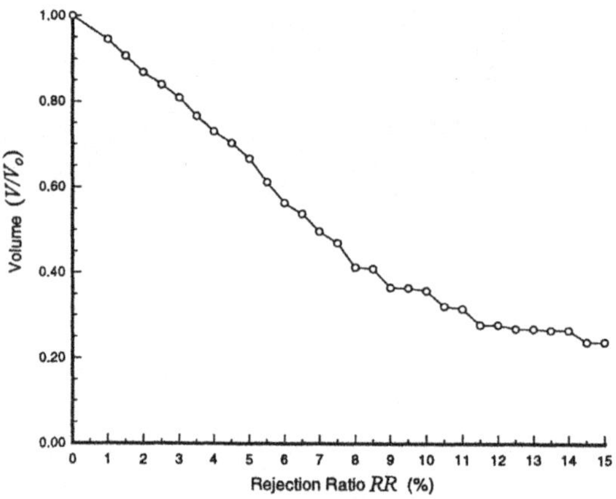

Fig. 2.10. History of volume reduction for a Michell type structure with two fixed supports.

Similar to the example of two-bar frame, we find that although there has been significant volume reduction, the increase in the maximum von Mises stress is relatively small. A comparison in terms of stresses and volume between the initial rectangular model in Figure 2.7 and the optimized design in Figure 2.8(c) is given in Table 2.2.

Table 2.2. Comparison between the initial rectangular model in Figure 2.7 and the optimized design in Figure 2.8(c)

	σ_{min}^{vm} (kPa)	σ_{max}^{vm} (kPa)	$\sigma_{min}^{vm} / \sigma_{max}^{vm}$	Volume (m^3)
Initial model	0.0130	33.4805	0.0004	5.000
Optimized design	5.8918	38.7176	0.1522	1.192

The ESO result in Figure 2.8(c) bears a strong resemblance to the original Michell truss in Figure 2.6. Both structures consist of an arch and several spokes between the load and the top of the arch, and the two legs of the arch are at an angle of 45° to the horizontal line.

However, there are also notable differences between the Michell solution and the ESO solution. There are obvious reasons for these differences. First, the Michell solution is based on truss assumption, i.e., all members are assumed to be pin-jointed to each other. But the ESO solution is obtained from a continuum model. Secondly, in the Michell solution all the members are assumed to be of a uniform cross-sectional area, while the ESO model has no such restriction. In fact the ESO solution indicates that it is desirable to have larger cross-sectional areas for the arch than for the spokes.

To further illustrate the nature of the stress based ESO technique, we apply the evolutionary procedure to a slightly modified model. Instead of putting fixed

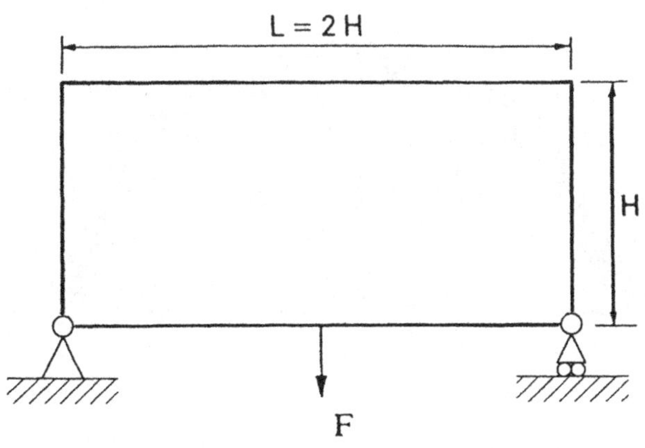

Fig. 2.11. Design domain for a Michell type structure with a roller support.

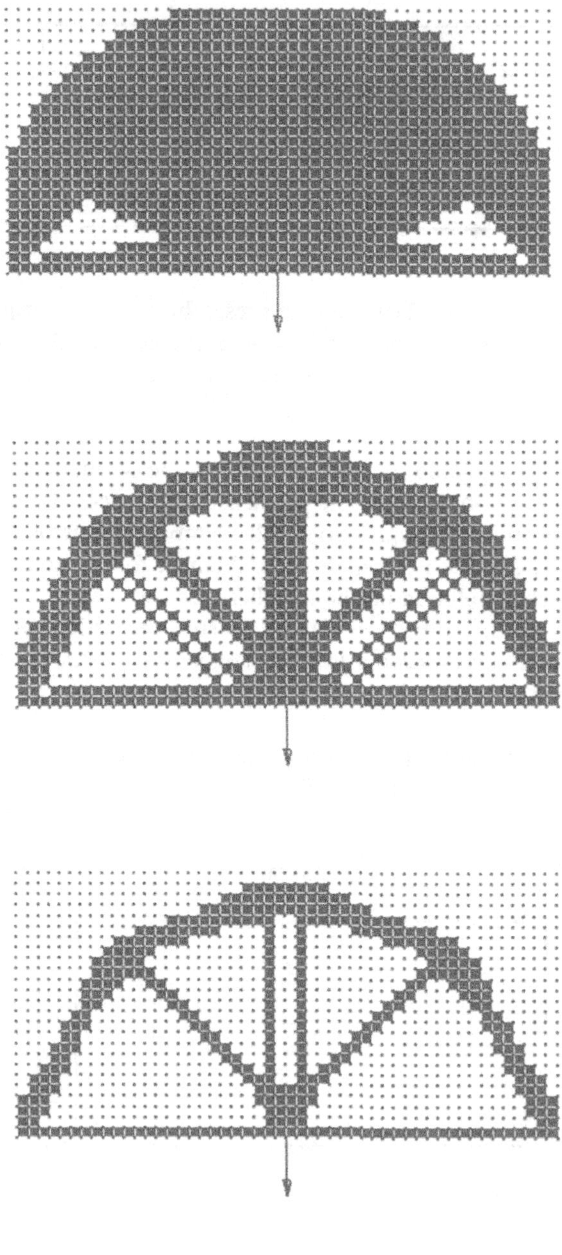

(a)

$RR = 5\%$

(b)

$RR = 10\%$

(c)

$RR = 15\%$

Fig. 2.12. ESO solutions for a Michell type structure with a roller support.

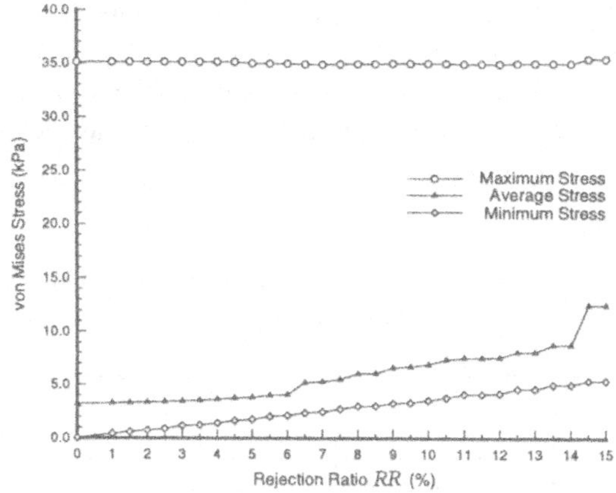

Fig. 2.13. Evolution histories of von Mises stress for a Michell type structure with a roller support.

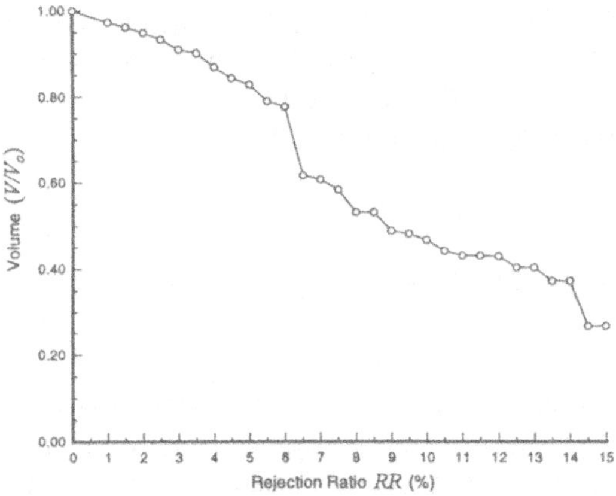

Fig. 2.14. History of volume reduction for a Michell type structure with a roller support.

supports at both bottom corners, the support on the right is now changed to a roller, as shown in Figure 2.11.

Keeping all other conditions and parameters the same as in the previous example, we now obtain a different set of designs shown in Figures 2.12(a-c). It is seen that two extra bars have appeared at the bottom, which are obviously in tension and trying to prevent the bottom right corner from moving further to the right. The evolution histories of the stresses are plotted in Figure 2.13 and the volume reduction history in Figure 2.14. A comparison of the stresses and volume is between the initial rectangular design in Figure 2.11 and the optimized design in Figure 2.12(c) is given in Table 2.3.

Table 2.3. Comparison between the initial rectangular model in Figure 2.11 and the optimized design in Figure 2.12(c)

	$\sigma_{\min}^{vm}$ (kPa)	$\sigma_{\max}^{vm}$ (kPa)	$\sigma_{\min}^{vm}/\sigma_{\max}^{vm}$	Volume (m^3)
Initial model	0.0159	35.1284	0.0005	5.000
Optimized design	5.3764	35.4071	0.1518	1.336

It is observed that throughout the evolutionary history, the maximum von Mises stress has hardly changed. For the optimized design shown in Figure 2.12(c), the volume has been reduced by 73% compared to the initial rectangular model, while the maximum stress has only increased by 0.8%.

2.5 Structures with Uniform Surface Stress

In the previous examples material can be removed from any part of a structure as long as the stress level in that part is relatively low compared to the maximum. This allows internal cavities being created and new topologies being generated.

However, in many design assignments internal cavities are not allowed and the designer can only modify the structural boundaries. These are traditionally classified as *shape optimization problems*.

In the previous examples the ESO method performs both shape and topology optimizations simultaneously. If we only allow material to be nibbled away from the boundaries, the ESO will be doing just shape optimization.

One interesting example of using the such an ESO method for shape optimization is given by Querin (1997), who seeks to find the optimal shape for an object

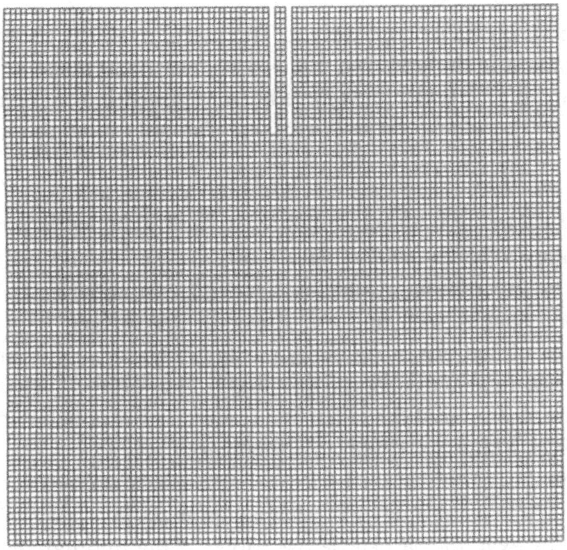

Fig. 2.15. Initial model for an object under gravity load.

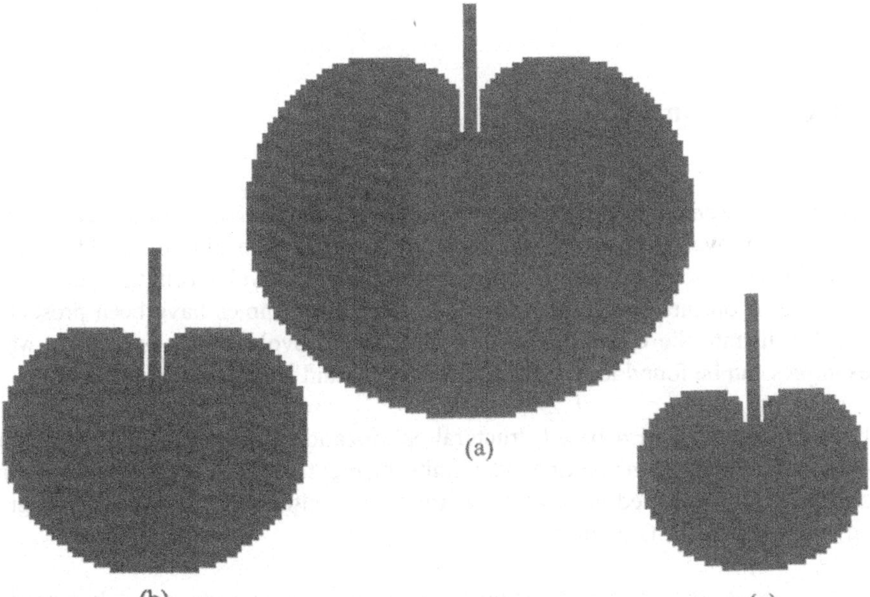

(a)

(b)

(c)

Fig. 2.16. ESO solutions for an object under gravity load: (a) steady state 15; (b) steady state 20; (c) steady state 25.

hanging in the air under its own weight. The initial model is shown in Figure 2.15. The square domain is divided into 100 × 100 four node plane stress elements. As shown in Figure 2.15, 48 elements have been removed in the initial model in order to create a stalk. The top of the stalk is fixed. The only loading on this object is the gravity.

With RR_0 and ER being both set to 0.05%, we apply the evolutionary procedure to the model. Figure 2.16 shows the shapes of the object at steady states 15, 20 and 25, respectively. By removing elements from the boundaries based on the stress level, we obtain shapes with uniform stress on the surface. The results in Figure 2.16 remind us of the shape of certain fruits such as plums and cherries. It is indicated that nature also evolves shapes that lead to uniform surface stress. Indeed this is the main objective of the extensive research of Mattheck and his co-workers (Mattheck 1990a; Mattheck 1990b; Mattheck *et al.* 1992).

This simple ESO technique for shape optimization may find many other applications. For example, it can be used to find the optimal shape of an internal cutout in a structure so that the stress on the surface of the cutout is uniform. Chapter 8 presents a more detailed discussion on the ESO technique for shape optimization.

2.6 Conclusion

We have presented in this chapter the simplest form of the ESO method. The material removal is based on the local stress level. The aim is to obtain more efficient designs by creating a more uniform stress level throughout the whole structure or on the surface of a structure. Several examples have been presented, which illustrate the capability of this stress based evolutionary procedure. More examples can be found in Xie and Steven (1993) and Hinton and Sienz (1995).

Unlike many other FEA based structural optimization techniques, the ESO method does not require re-generating new finite element meshes even when the final structure has departed substantially from the initial design. This is a great advantage of the ESO method.

FEA computer programs and commercial packages are nowadays widely available. One such FEA code is STRAND6 (G+D Computing 1993), which has been used as a FEA computational engine for this work. Even with limited access to the source code, the ESO can be implemented readily. After a finite element analysis, stresses of each element are normally stored in a file. From this file the von Mises stress can be read or calculated for each element. By comparing the von Mises stresses of all elements, one can immediately decide which elements are to be eliminated from the model. The element removal can be done by simply assigning the material property number of the rejected elements to zero and ignore these elements when the global stiffness matrix is assembled in subsequent stress analyses. As more and more elements are removed, the number of equations to be solved at each iteration becomes less and less.

Another way of 'removing' an element is to reduce its elastic modulus or dimensions such as element thickness to a very small value. For example, Hinton and Sienz (1995) reduce the elastic modulus of the elements to be removed by a factor of 10^5. In doing so, however, due to the large difference in elastic moduli or thicknesses among elements the global stiffness matrix may become ill-conditioned, which in turn may lead to grave errors in the numerical solutions of displacements and stresses. Also no matter how many elements are "removed" by reducing the elastic modulus or the plate thickness, the number of equations to be solved for each stress analysis remains the same. For these reasons we prefer to use the option of totally eliminating elements from the finite element mesh.

2.7 References

G+D Computing (1993) *STRAND6 Reference Manual and User Guide*, Sydney.

Hemp, W.S. (1973) *Optimum Structures*, Clarendon Press, Oxford.

Hinton, E. and Sienz, J. (1995) Fully stressed topological design of structures using an evolutionary procedure. *Eng. Computations* **12**, 229-244.

Mattheck, C. (1990a) Engineering components grow like trees. Materialwissesch. Werkstofftech. **21**, 143-168.

Mattheck, C. (1990b) Design and growth rules for biological structures and their application to engineering. *Fatigue Fract. Eng. Mater. Struct.*, **13**(5), 535-550.

Mattheck, C. Erb, D., Bethge, K. and Begemann, U. (1992) Three dimensional shape optimization of a bar with a rectangular hole. *Fatigue Fract. Eng. Mater. Struct.* **15**(4), 347-351.

Michell, A.G.M. (1904) The limits of economy of material in frame-structures. *Phil. Mag.* **8**, 589-597.

Querin, O.M. (1997) *Evolutionary Structural Optimization, Formulation and Implementation.* PhD thesis, Department of Aeronautical Engineering, University of Sydney.

Xie, Y.M. and Steven, G.P. (1993) A simple evolutionary procedure for structural optimization. *Comput. Struct.* **49**, 885-896.

Chapter 3 ESO for Multiple Load Cases and Multiple Support Environments

Many structures are subjected to different load cases at different times, and some structures are held or supported in different ways at different times. The former is referred to as structures with *multiple load cases*, and the later *multiple support environments.* The ESO procedures for the optimal design of structures with multiple load cases and multiple support environments are the topics of this chapter. Original research on these topics was first reported by Xie and Steven (1994) and Steven *et al.* (1995).

3.1 Introduction

In reality most structures are required to operate under multiple load cases. Each of these load cases acts on the structure at a different time. For example, a bicycle is under a certain set of forces when the rider is braking, but the magnitudes and directions of the forces on the bicycle will be totally different when the rider starts to accelerate. Of course a bicycle has to be designed to account for both of these, and more, load cases.

Equally as important as having multiple load cases is to allow the structure to be held or supported in different ways at different times of its operation, i.e. a structure with multiple support environments. For example, an aircraft is supported by distributed air pressure when in flight but at discrete points by the undercarriage when on the ground. As another example, a car body is supported at the suspension in normal driving but for jacking there could be different support conditions for a repair shop hydraulic jack or a scissors type jack.

In this chapter the ESO method is extended to the optimal design of structures with multiple load cases and multiple support environments. Similar to the previous chapter, the material removal criterion is based on the local stress level and the objective is to achieve more uniform stress distributions in the new designs.

3.2 ESO for Multiple Load Cases

The finite element analysis for a structure with multiple load cases is not much more expensive than a single load case, because the global stiffness matrix needs only to be decomposed once for each given mesh, no matter how many different load cases are set up for the model. The simple ESO procedure presented in Chapter 2 can be easily extended to the optimal design of structures with multiple load cases.

After the finite element analysis, a stress distribution is obtained for each load case. The rejection criterion (2.2) is then checked for each load case. An element is eliminated from the structure only if this rejection criterion is satisfied by all load cases. The final structure is an optimal design in the sense that every part of the remaining material has its own role to play in at least one load case and possibly in all load cases.

Similar to a multiple load case situation, when structures such as bridges are subjected to moving load, the force acting on a structure moves from one part to the other. A moving load can be conveniently approximated to multiple load cases, by applying this load sequentially to a finite number of locations along the path of the moving load.

3.3 ESO for Multiple Support Environments

When a structure is held or supported in several different ways, the same FEA mesh can be used for all support environments. However, for each support environment the global stiffness matrix is different and needs to be solved separated. Therefore the computational cost for an extra support environment is much higher than that for an extra load case.

After the stress distributions have been found for all support environments the rejection criterion (2.2) is checked for each support environment. An element is removed from the structure only if this rejection criterion is satisfied by all support environments. The remaining material in the final design will be such that it is required by at least one support environment and possibly by all support environments.

3.4 Examples of ESO for Multiple Load Cases

3.4.1 A Bearing Pedestal

In this example an optimal design for the bearing pedestal shown in Figure 3.1 is sought. The material on both sides of the structure provides a basis for the attachment of the pedestal to the underlying surface by, for example, bolt joints. An optimization model is set in such a way that these parts are not to be removed or changed during the evolutionary process. Furthermore a minimum thickness of material surrounding the bearing hole is required. These regions constitute the so-called non-design domain. Material properties of these regions can be the same or different from those of other regions, but elements in the non-design domain are not allowed to be removed even the stress level is low. In the finite element implementation, we use a different material number for elements in the non-design domain. Elements with such a special material number will never be removed from the structure.

In addition to the bolt joint loads which are present at all time, there is a load acting on the bearing from a shaft. This load is acting either upwards or downwards at different time. The magnitude of the upward force is twice as big as that of the downward force.

It is assumed that the pedestal is in full contact with the underlying surface. Due to symmetry of the geometry and loading, only the right half of the structure is modelled. Figure 3.2 shows the finite element mesh of three node triangles produced by an automatic mesh generator developed by Peraire *et al.* (1987).

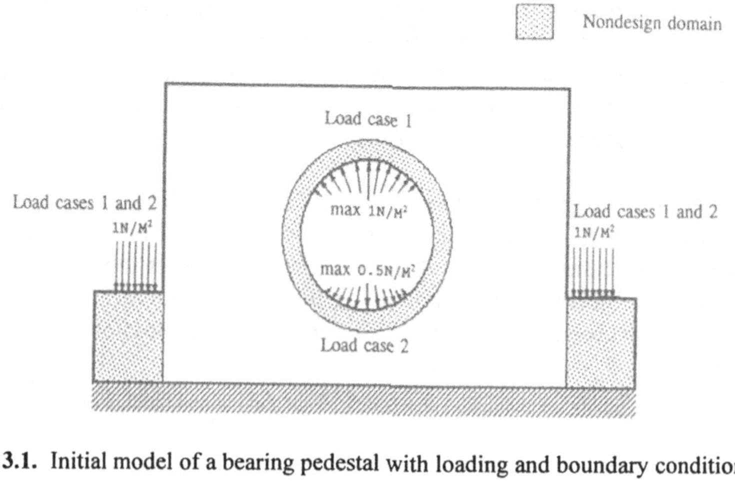

Fig. 3.1. Initial model of a bearing pedestal with loading and boundary conditions.

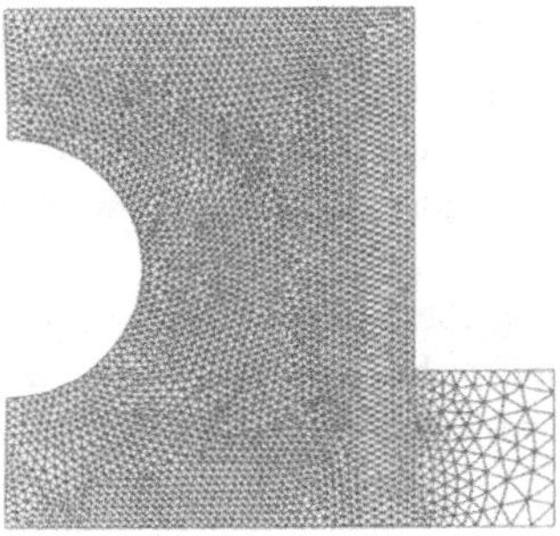

Fig. 3.2. Finite element mesh for the bearing pedestal.

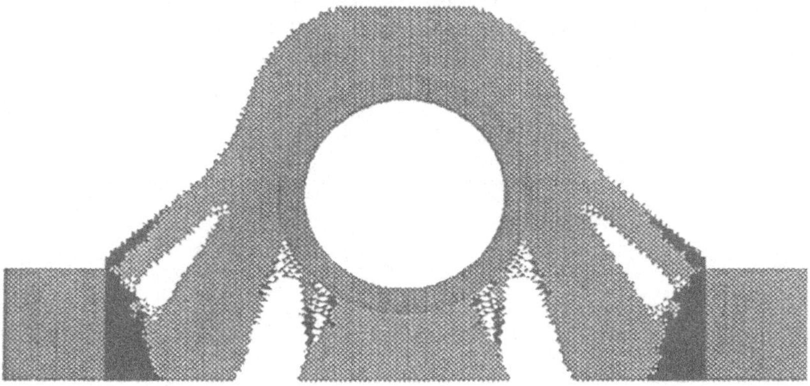

Fig. 3.3. Optimal design for the bearing pedestal with two load cases.

By considering the two load cases shown in Figure 3.1, we start the ESO process with the initial rejection ratio RR_0 = 1% and the evolutionary rate ER = 1%. Plane stress conditions are assumed, with Young's modulus E = 200 GPa and Poisson's ratio ν = 0.3. Figure 3.3 shows the ESO result when the rejection ratio RR reaches 10%. The lowly stressed square corners in the original model are removed and a smooth outer surface is obtained. Furthermore, internal holes are created and in doing so an extra large portion of unneeded material is removed from the structure. Such a change in the topology is of course impossible in the usual boundary variation methods. In Figure 3.3 there are regions at the edges of the hole that are not fully populated with elements, which implies that thinner material could be used for these regions.

In Figure 3.3 the darker regions represent material which is required by both load cases. The trapezoidal part below the bearing hole is required by load case 2 only. The rest of the grey regions (except the non-design domain) are required by load case 1 only. For a clearer illustration, the reader is referred to Xie and Steven (1994) where the result in Figure 3.3 is presented in colours.

It is interesting to note that the ESO solution shown in Figure 3.3 is close to what Rasmussen *et al.* (1992) has obtained in a similar example using the homogenization method followed by a traditional shape optimization.

3.4.2 A Bridge with a Moving Load

Figure 3.4 shows the initial configuration of a bridge. For a structure such as the bridge the most common loading is a traffic load, which moves from one part of the surface to the other. This continuous moving load can be approximated to a finite number of load cases. Each of the load cases is applied at a different time but with the same magnitude as the traffic load itself.

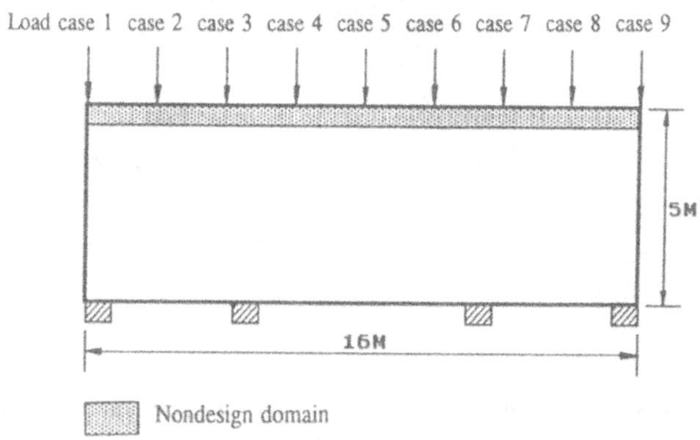

Fig. 3.4. Initial model of a bridge with loading and boundary conditions.

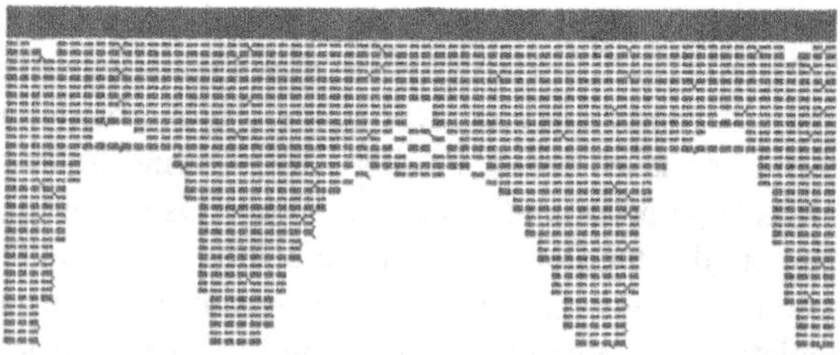

Fig. 3.5. Optimal design for the bridge with nine load cases.

In this example, a point load is moving from the left to the right of the bridge on the top surface. It is approximated to nine load cases with an equal distance between each other, as shown in Figure 3.4. The bridge has four supports at the bottom. The body of the bridge is discretized into 60 x 40 four node plane stress elements. In this model the deck on the top of the bridge is not allowed to be removed and is therefore defined as non-design domain (0.5m deep). Although the structure and the boundary conditions are symmetrical, the full structure has to be analysed because the loading is asymmetric except for load case 5.

Taking into account all the nine load cases, the ESO method is applied to this model with the initial rejection ratio $RR_0 = 1\%$ and the evolutionary rate $ER = 1\%$. When the rejection ratio RR reaches 10%, we obtain the result shown in Figure 3.5. This result bears a striking resemblance to classical arch bridges.

3.4.3 Design of Bicycle Frames

One of our ongoing research projects is the application of the ESO technique to the optimal design of bicycle frames. With the advance in composite materials and the level of competition among professional riders, new designs of bicycles have received ever increasing attention from designers, who no longer restrict themselves to the traditional diamond-shaped frame (with steel tubes linking the handlebar, saddle, chain ring and rear wheel). The best example of this new development is the LotusSport bike (Thisdell 1992), with which Britain's Chris Boardman won a gold medal at the Olympic Games in Barcelona. The entire frame of the LotusSport bike is moulded from a single piece of carbon fibre material.

By applying the ESO technique, we can now explore the possibilities of finding optimal designs for bicycle frames. Figure 3.6 shows one of the 2D results obtained using ESO (Mendis 1996). In this model four load cases which represent loading conditions for starting, speeding, rolling and braking are included. The remaining material in Figure 3.6 is required by at least one of the four load cases.

In order to reduce aerodynamic drag force, it is desirable to avoid internal holes in a bicycle frame. This leads to an attempt to produce a monocoque design using carbon fibre material. The nibbling option of ESO is applied, where material can only be removed from the boundary. Figure 3.7 shows the shape of a monocoque bicycle frame obtained from ESO for the structural environment of an 80 kg rider doing starting, speeding, rolling and braking. The computer generated shape has been made from carbon fibre and then fitted with off-the-shelf components including the front fork, handlebar, saddle, wheels, chain ring and cog.

Fig. 3.6. Optimal design for a bicycle frame with five load cases.

Fig. 3.7. A carbon fibre monocoque bicycle frame after final assembly.

The carbon fibre frame itself weighs only 1.2 kg. Details about the ESO bicycle shown in Figure 3.7 can be found in Medved (1995). Static and on-road tests are being carried out to further assess the performance of this bicycle and to improve on the current design.

3.5 Examples of ESO for Multiple Support Environments

3.5.1 Fixed and Rolling Arch Supports

In Chapter 2 the optimal arches for two Michell type structures have been presented, one for fixed supports and the other for a roller support. The real support conditions at the arch abutments are not very certain. Such an uncertainty could be very dangerous from the viewpoint of a structural designer. For example, the optimal arch design for fixed supports shown in Figure 2.8 would immediately collapse if one of the supports is actually (or later becomes) a roller. In these circumstances a safer approach is to find an optimal design which can accommodate all possible support environments.

In this example, we seek the most fully stressed design for the arch which may operate safely in both fixed and roller support conditions. Using the ESO procedure for multiple support environments as described in Section 3.3 we obtain the solutions shown in Figure 3.8. These can be compared with the solutions in Figures 2.8 and 2.12 which are obtained for individual support environments.

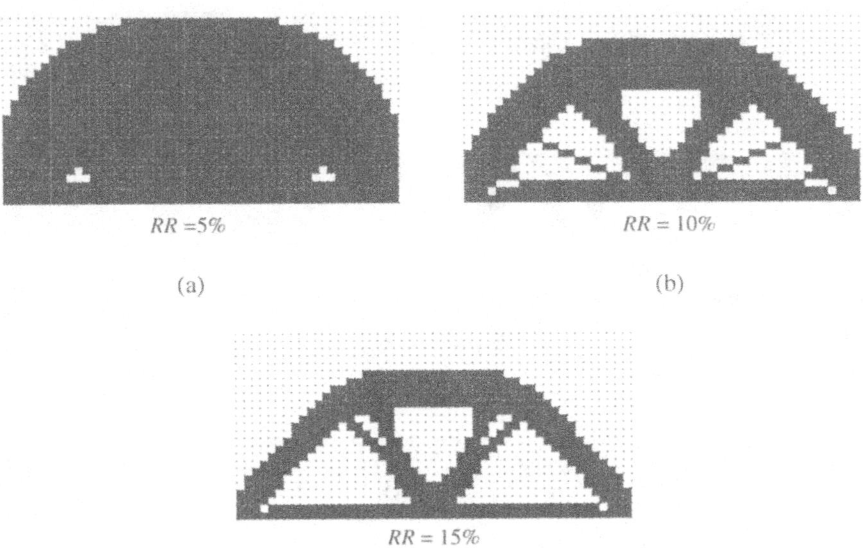

Fig. 3.8. ESO solutions for an arch structure with two support environments.

3.5.2 Sinking Bridge Support

Consider the bridge example in Figure 3.4 again. It could arise that the bridge may have one of its four supports in a river bed, and the design needs to allow for the possibility of that support sinking or being swept away after a collision with a boat. We now apply the ESO to a new bridge model, assuming two support environments: one with all four supports, and the other with three supports (the second support from the left is eliminated).

Figure 3.9 shows the evolving structure as the ESO is trying to find the most fully stressed design for the bridge when taking into account the support environments. It is seen that at the beginning of the evolution the design makes some use of the vulnerable support but towards the end it gives up any reliance upon this support.

Note that the finite element mesh used in this example is finer than that in Section 3.4.2, and the initial rejection ratio $RR_0 = 1\%$ and the evolutionary rate $ER = 0.5\%$ are assumed.

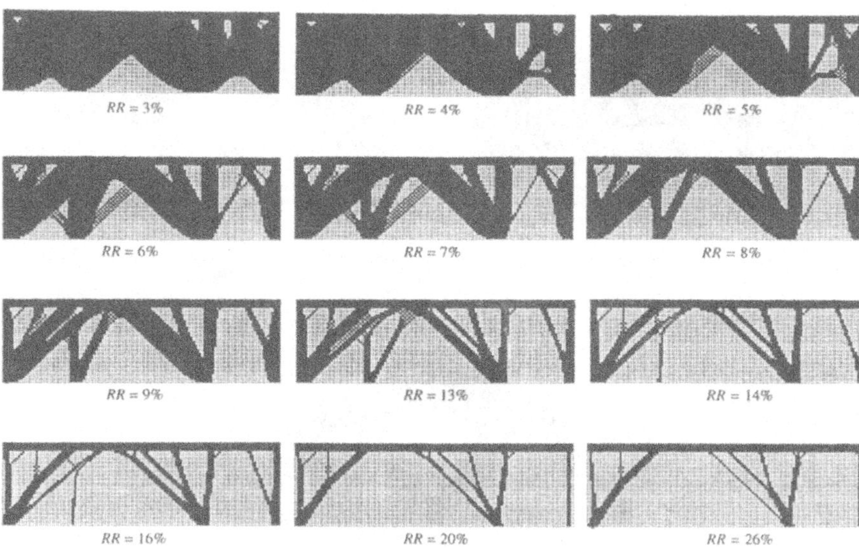

Fig. 3.9. Evolving structural shape for a bridge with a vulnerable support.

3.6 Conclusion

The working environment of a real structure may contain many different loading and support conditions. The ESO method presented in this chapter can be used to find the most fully stressed design for the structure when all load cases and support environments are considered. For this purpose a material removal criterion based on stress level is used.

For other design objectives such as maximizing stiffness, frequency or buckling load, different material removal criteria, which are presented in subsequent chapters, have to be used. However, no matter what rejection criterion is used, the same concept applies: in the final design every part of the remaining material has its own role to play in at least one, and possibly all, load cases and support environments.

3.7 References

Medved, J.J. (1995) *Design and Optimization of a Carbon Monocoque Bicycle Frame.* Fourth Year Thesis, Department of Aeronautical Engineering, University of Sydney.

Mendis, B.B.C. (1996) *Optimization of a Carbon Composite Bicycle Frame.* Fourth Year Thesis, Department of Aeronautical Engineering, University of Sydney.

Peraire, J., Vahdati, M., Morgan, K. and Zienkiewicz, O.C. (1987) Adaptive remeshing for compressible flow computations. *J. Comp. Phys.* **72**, 449-466.

Rasmussen, J., Lund, E. and Birker, T. (1992) Collection of examples: CAOS optimization system, 3rd edition, *Special Report No. 13*, Institute of Mechanical Engineering, Aalborg University , Denmark.

Steven, G.P., Querin, O.M. and Xie, Y.M. (1995) Multiple constraint environments for evolutionary structural optimization. In: N. Olhoff and G.I.N. Rozvany (eds.), *Proceedings of the First World Congress on Structural and Multidisciplinary Optimization*, Goslar, Germany, May 28 - June 2, 1995, Pergamon, Oxford, 213-218.

Thisdell, D. (1992) For sale: the world's fastest bicycle. *New Scientist* **135**, No. 1838, 23-24.

Xie, Y.M. and Steven, G.P. (1994) Optimal design of multiple load case structures using an evolutionary procedure. *Eng. Computations* **11**, 295-302.

Chapter 4 Structures with Stiffness or Displacement Constraints

Stiffness is one of the key factors that need to be taken into account in the design of structures such as buildings and bridges. It is often required that a structure be stiff enough so that the maximum deflection is within a prescribed limit. This chapter describes the evolutionary structural optimization procedures for the simultaneous shape and layout optimization of structures with stiffness or displacement constraints. Much of the material in this chapter derives from the work by Chu *et al.* (1996a-d).

4.1 Overall Stiffness Constraint

In finite element analysis (FEA), the static behaviour of a structure is represented by the following equilibrium equation

$$[K]\{u\} = \{P\} \tag{4.1}$$

where $[K]$ is the global stiffness matrix, $\{u\}$ is the nodal displacement vector and $\{P\}$ is the nodal load vector.

The strain energy of the structure, which is defined as

$$C = \frac{1}{2}\{P\}^{\mathrm{T}}\{u\}, \tag{4.2}$$

is commonly used as the inverse measure of the overall stiffness of the structure. C is also known as the mean *compliance*. It is obvious that maximizing the overall stiffness is the same as minimizing the strain energy.

Consider the removal of the ith element from a structure comprising n finite elements. The stiffness matrix will change by

$$\Delta[K] = [K^*] - [K] = - [K^i], \tag{4.3}$$

where $[K^*]$ is the stiffness matrix of the resulting structure after the element removal and $[K^i]$ is the stiffness matrix of the ith element. It is assumed that the removal of the element has no effect on the load vector $\{P\}$. By ignoring a higher order term, we obtain the change of the displacement vector from equation (4.1) as

$$\{\Delta u\} = - [K]^{-1}[\Delta K]\{u\}. \tag{4.4}$$

From equations (4.3) and (4.4) we have

$$\Delta C = \frac{1}{2}\{P\}^{\mathrm{T}}\{\Delta u\} = - \frac{1}{2}\{P\}^{\mathrm{T}}[K]^{-1}[\Delta K]\{u\} = \frac{1}{2}\{u^i\}^{\mathrm{T}}[K^i]\{u^i\}, \tag{4.5}$$

where $\{u^i\}$ is the displacement vector of the ith element. We thus define

$$\alpha_i = \frac{1}{2}\{u^i\}^{\mathrm{T}}[K^i]\{u^i\} \tag{4.6}$$

as the sensitivity number for problems with an overall stiffness constraint. This sensitivity number indicates the change in the strain energy as a result of removing the ith element. In fact α_i is the element strain energy and it can be easily calculated at the element level using the element stiffness matrix and the displacement vector of the element. It is worth pointing out that both C and α_i always have positive values.

The optimization objective is to find the lightest structure while satisfying the stiffness constraint, typically given in the form

$$C \le C^*, \tag{4.7}$$

where C^* is a prescribed limit for C. In general, when an element is removed, the overall stiffness of a structure reduces and correspondingly the strain energy C increases. To achieve the optimization objective through element removal, it is obviously most effective to remove the element which has the lowest value of α_i so that the increase in C is minimum.

The evolutionary procedure for optimization with an overall stiffness constraint can be summarized as follows:

Step 1: Discretize the structure using a fine mesh of finite elements;
Step 2: Solve the static equilibrium equation (4.1) for the given load;
Step 3: Calculate the sensitivity number for each element using equation (4.6);
Step 4: Remove a number of elements which have the lowest sensitivity numbers;

Step 5: Repeat Steps 2 to 4 until condition (4.7) can no longer be satisfied.

The number of elements to be removed at each iteration is determined by a prescribed *element removal ratio*, which is defined as the ratio of the number of elements to be removed at each iteration to the total number of elements in the initial or the current FEA model. Typical values for the element removal ratio are 1% and 2%. Further discussions on the influence of the amount of material removed at each iteration on the final solutions can be found in Chu *et al.* (1996b). With a prescribed element removal ratio, the number of elements to be removed in Step 4 is equal to the removal ratio times the number of elements in the initial or the current FEA model, rounded off to the nearest integer. In the case where the symmetry of a structure needs to be maintained, an even number of elements should be removed at each iteration. In Step 5 the evolutionary procedure can also be terminated when a prescribed percentage of volume has been eliminated from the structure.

4.2 Displacement Constraint

It is often required that the maximum displacement of a structure or the displacement at a specific location should be within a prescribed limit. If a limit is imposed on the jth displacement component u_j, the displacement constraint may be given in the form

$$|u_j| \leq u_j^*, \tag{4.8}$$

where u_j^* is the prescribed limit for $|u_j|$. To find the change in u_j due to an element removal, we introduce an unit load vector $\{F^j\}$, in which only the jth component is equal to unity and all the others are equal to zero. Multiplying equation (4.4) by $\{F^j\}^T$ we obtain

$$\Delta u_j = \{F^j\}^T [K]^{-1} [K^i]\{u\} = \{u^j\}^T [K^i]\{u\} = \{u^{ij}\}^T [K^i]\{u^i\}, \tag{4.9}$$

where $\{u^j\}$ is the displacement due to the unit load $\{F^j\}$, $\{u^i\}$ and $\{u^{ij}\}$ are the element displacement vectors containing the entries of $\{u\}$ and $\{u^j\}$, respectively, which are related to the ith element. The calculation of the right hand side of equation (4.9) can be done at element level, as all the entries in $[K^i]$, $\{u^i\}$ and $\{u^{ij}\}$ are zero except for those which are related to the ith element. The value

$$\alpha_{ij} = \{u^{ij}\}^T [K^i]\{u^i\} \tag{4.10}$$

indicates the change in the specified displacement component u_j due to the removal of the ith element. It should be noted that, unlike α_i in equation (4.6), α_{ij} can be either positive or negative, which implies that u_j may change in opposite directions. To minimize the change in $|u_j|$, it is obviously most effective to remove the element whose α_{ij} is closest to zero. We thus define

$$\alpha_i = |\alpha_{ij}| \tag{4.11}$$

as the sensitivity number for problems with a displacement constraint. This sensitivity number can be easily calculated from the finite element solutions of the static problem (4.1) for both the actual load $\{P\}$ and the unit load $\{F^j\}$.

The evolutionary procedure for optimization with a displacement constraint is given as follows:

Step 1: Discretize the structure using a fine mesh of finite elements;
Step 2: Solve the static equilibrium equation (4.1) for the given load $\{P\}$ and the virtual unit load $\{F^j\}$;
Step 3: Calculate the sensitivity number for each element using equation (4.11);
Step 4: Remove a number of elements which have the lowest sensitivity numbers;
Step 5: Repeat Steps 2 to 4 until condition (4.8) can no longer be satisfied.

For the special case where there is only one point load Q acting on a structure and a constraint is imposed on the displacement at the same location and in the same direction as the load, $\{u^j\}$ is equal to $(1/Q)\{u\}$ and therefore equation (4.1) needs not to be solved for the unit load $\{F^j\}$. The sensitivity number for this case differs only by a constant factor $(2/Q)$ from that for the problem with an overall stiffness constraint as discussed in the previous section. Such a constant factor does not change the relative order of the sensitivity numbers of the elements.

4.3 Examples of Optimization with a Displacement Constraint

4.3.1 A Short Cantilever

The cantilever beam shown in Figure 4.1 is under plane stress conditions. The left hand side of the beam is fixed. A vertical load of 3 kN is applied at the middle of the free end. The dimensions of the beam are $L_x = 0.16$ m, $L_y = 0.10$ m and the thickness $t = 0.001$ m. The Young's modulus $E = 207$ GPa and the Poisson's ratio $\nu = 0.3$ are assumed. The initial vertical displacement at the middle of the free end is 0.33 mm.

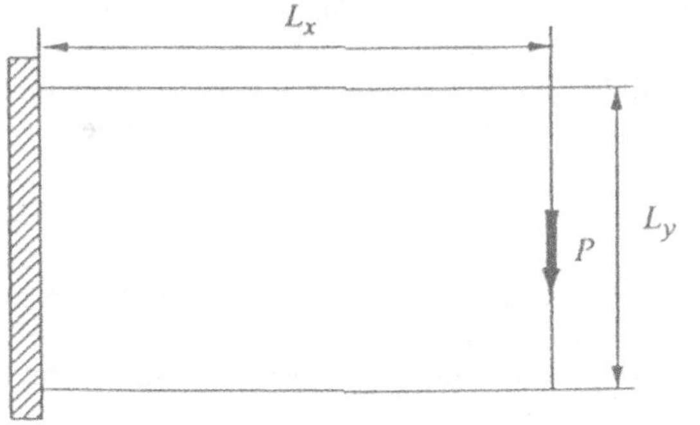

Fig. 4.1. Design domain for a short cantilever.

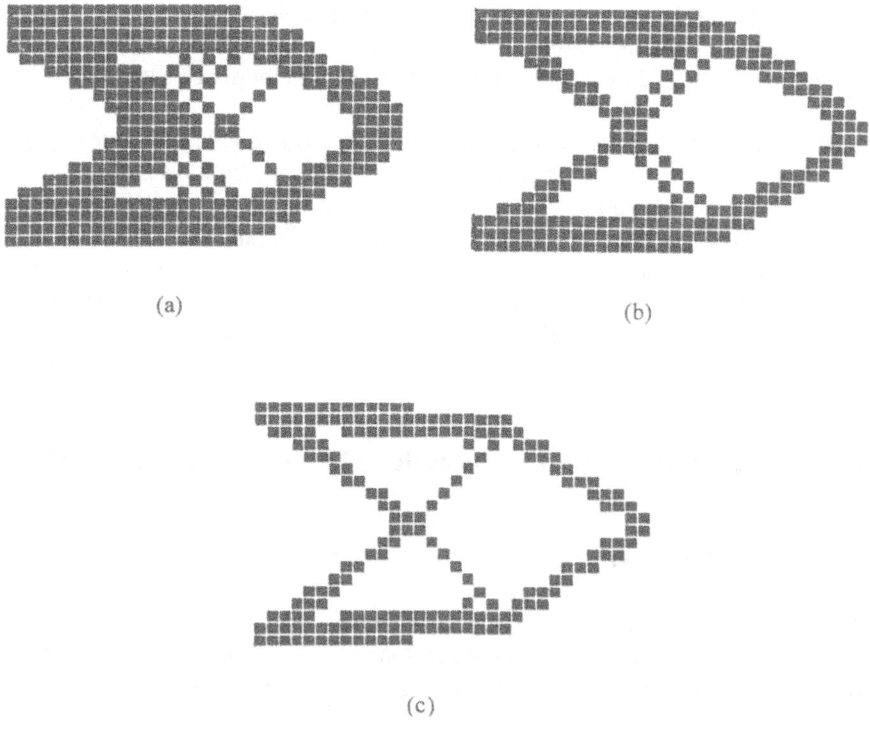

Fig. 4.2. Optimal designs for the short cantilever for limits: (a) 0.50 mm; (b) 0.75 mm; (c) 1.00 mm.

The design domain is divided into 32 × 20 four node elements. At each iteration the number of elements removed from the structure is chosen to be 12, which corresponds to 2% of the initial elements rounded off to the nearest even number. An even number is used so that all the new designs remain symmetrical.

A displacement constraint is imposed on the vertical displacement at the middle of the free end. The evolutionary procedure gives new designs for displacement limits of 0.50, 0.75 and 1.00 mm, respectively, as shown in Figure 4.2. The volumes of these designs are 56.87, 39.37 and 30.00%, respectively, of the initial volume.

One way of measuring the efficiency of a new design is to compare its volume with that of the *optimized uniform design,* which is obtained by simply reducing the thickness of the initial design until the displacement with the constraint reaches the prescribed limit. In this example, for the same given limits of 0.50, 0.75 and 1.00 mm, the volumes of the optimized uniform designs are 66, 44 and 33% of the initial volume, respectively. In comparison with the optimized uniform design, the ESO designs in Figure 4.2 have resulted in volume reductions of 13.8, 10.5 and 9.0%, respectively.

4.3.2 A Michell Type Structure

In this example we apply the ESO procedure to a Michell type structure (Michell 1904; Hemp 1973). As shown in Figure 4.3 an optimum structure is to be found to transfer a vertical force P to the circular fixed support. The dimensions of the rectangular domain are 0.55 m by 0.4 m. The thickness of the plate is 0.001 m. The radius of circular fixed support is equal to 0.1m. The Young's modulus E = 205 GPa and the Poisson's ratio $\nu = 0.3$ are used. The structure is assumed to be

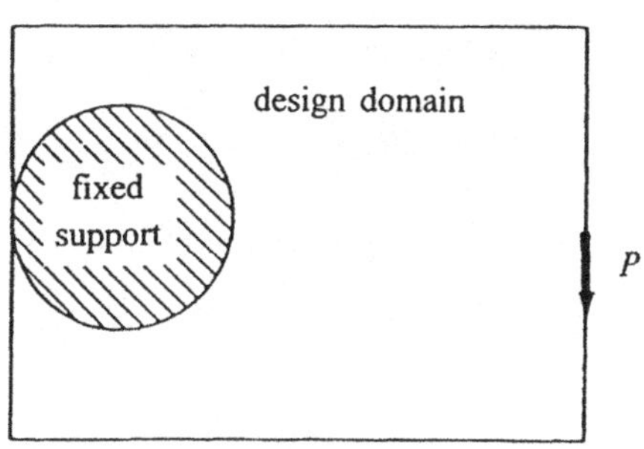

Fig. 4.3. Design domain for a Michell type structure.

under plane stress conditions. With the applied force $P = 50$ kN, the initial displacement of the loaded point in the vertical direction is 2.87 mm.

Due to symmetry, only the top half of the model is analyzed using 110×40 four node elements. The circular support is represented by fixing the nodes closest to the circle. Elements within the circle are not considered in the solution process. A displacement constraint is imposed on the vertical displacement at the loaded point. At each iteration the number of element removed is 44, which corresponds to 1% of the initial elements. The solutions for the displacement limits of 5, 7 and 9 mm are given in Figure 4.4, which have the volumes of 47.45, 36.26 and 29.26%, respectively, of the initial volume. The volumes of the corresponding optimized uniform designs are 57.5, 41.0 and 31.9% of the initial volume, which are significantly higher than the volumes of the ESO designs in Figure 4.4.

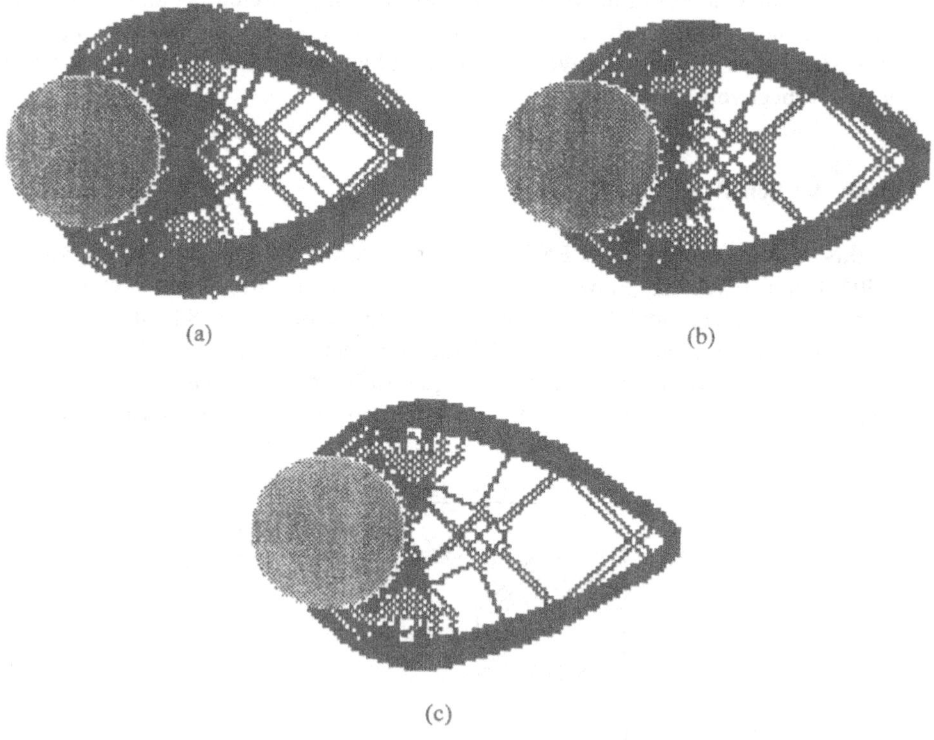

Fig. 4.4. Optimal designs for the Michell type structure for limits: (a) 5 mm; (b) 7 mm; (c) 9 mm.

4.3.3 The MBB Beam

The optimization of a support beam from a civil aircraft produced by Messerschmitt-Bölkow-Blohm has now become a classical problem in topology optimization (e.g. Olhoff *et al.* 1991; Zhou and Rozvany 1991; Jog *et al.* 1994). The MBB beam, as it is known, has the function of carrying the floor in the fuselage of an Airbus passenger carrier. The beam is 2400 mm long and 400 mm deep with a point load of 20 kN acting at the middle. The initial design of the beam is shown in Figure 4.5. It is required that the maximum deflection of the beam should not exceed 9.4 mm under the given load.

Because of symmetry, only half of the structure is modelled with 60 × 20 four node elements. The Young's modulus $E = 200$ GPa and Poisson's ratio $\nu = 0.3$ are assumed. Under plane stress conditions, the initial maximum displacement is 6.3 mm. Figure 4.6 shows the ESO design of the MBB beam for the displacement limit of 9.4 mm. In this example a removal ratio of 1% of the current number of elements is used. The volume of the ESO design is equal to 50.33% of the initial volume. The volume of the corresponding optimized uniform design is 67% of the initial volume. So the ESO design has a volume reduction of 24.88% compared to the optimized uniform design.

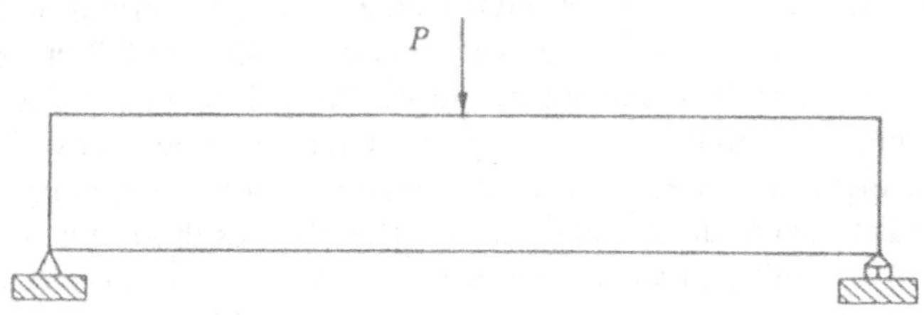

Fig. 4.5. Design domain for the MBB beam.

Fig. 4.6. Optimal design for the MBB beam.

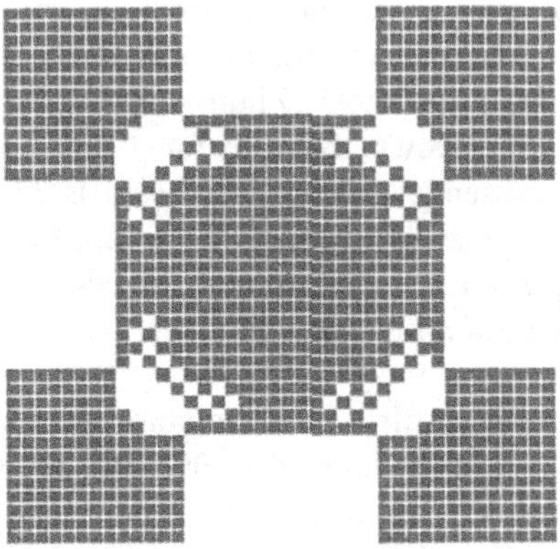

Fig. 4.7. Optimal design for a plate in bending.

4.3.4 A Plate in Bending

A square plate (0.20m × 0.20m × 0.0001m) is simply supported on the four edges and is loaded at the centre by a point load P = 0.04 N normal to its plane. The Young's modulus E = 174.7 GPa and Poisson's ratio ν = 0.3 are assumed. The initial out-of-plane displacement at the centre is 1.16 mm. The limit for this displacement is 1.6 mm. One quarter of the plate is modelled with 20 × 20 four node plate elements. At each iteration the number of element removed is 8, which corresponds to 2% of the initial elements. The ESO result for this problem is given in Figure 4.7. Similar to what have been reported in Atrek (1989) and Tenek and Hagiwara (1994), we find that hinge lines are formed between the central part and the four corners.

4.4 Constraint on the Difference of Two Displacements

In the design of building type structures it is often required that the relative displacement of adjacent floors be within a certain limit. This is one of the examples where a constraint is imposed on the difference of two displacements, which may be given in the form

$$|u_j - u_k| \le \delta_{jk}^{*}, \tag{4.12}$$

where δ^*_{jk} is a prescribed limit for the difference between the two displacement components u_j and u_k. The effect of removing the ith element on this displacement difference can be calculated as follows

$$\Delta|u_j - u_k| = |\{u^{ij}\}^T[K^i]\{u^i\} - \{u^{ik}\}^T[K^i]\{u^i\}| = |\{\delta^i_{jk}\}^T[K^i]\{u^i\}|, \tag{4.13}$$

where $\{u^{ij}\}$ and $\{u^{ik}\}$ are the element displacement vectors of the ith element due to a unit load applied at the location and direction of u_j and u_k, respectively, and

$$\delta^i_{jk} = \{u^{ij}\} - \{u^{ik}\}. \tag{4.14}$$

We thus define

$$\alpha_i = |\{\delta^i_{jk}\}^T[K^i]\{u^i\}| \tag{4.15}$$

as the sensitivity number for problems with a constraint on the difference of two displacements. To minimize the change in the displacement difference, it is obviously most effective to remove the element whose α_i from equation (4.15) is closest to zero.

The evolutionary procedure for optimization with a constraint on the difference of two displacements is given as follows:

Step 1: Discretize the structure using a fine mesh of finite elements;

Step 2: Solve the static equilibrium equation (4.1) for the given load $\{P\}$ and the two virtual unit loads corresponding to u_j and u_k;

Step 3: Calculate the sensitivity number for each element using equation (4.15);

Step 4: Remove a number of elements which have the lowest sensitivity numbers;

Step 5: Repeat Steps 2 to 4 until condition (4.12) can no longer be satisfied.

4.5 Multiple Displacement Constraints

For some structures it may be required that displacements at several different locations or directions be within prescribed limits. Such multiple displacement constraints may be given in the form

$$|u_j| \le u^*_j, \qquad (j = 1, m), \tag{4.16}$$

where m is the total number of displacements with constraints.

There are various ways of dealing with multiple displacement constraints in the literature such as using Lagrangian multipliers (e.g. Venkayya *et al.* 1973). A simpler approach is to use the weighted average of the expected changes in the displacements with constraints due to element removal. The weighted average may be given in the form

$$\alpha_i = \sum_{j=1}^{m} \lambda_j |\alpha_{ij}| \tag{4.17}$$

and defined as the sensitivity number for the ith element. In the above equation α_{ij} is calculated using equation (4.10) for each displacement constraint and the weighting parameter λ_j is given as

$$\lambda_j = |u_j| / u_j^*. \tag{4.18}$$

When a displacement is far below the prescribed limit, the corresponding weighting parameter will be very small, and thus the corresponding constraint will be of little significance in equation (4.17). However, it is possible that this constraint may soon become dominant when, after a few iterations, the corresponding displacement is approaching its limit.

The evolutionary procedure for optimization with multiple displacement constraints is given as follows:

Step 1: Discretize the structure using a fine mesh of finite elements;

Step 2: Solve the static equilibrium equation (4.1) for the given load $\{P\}$ and the virtual unit loads corresponding to all the displacement constraints;

Step 3: Calculate the sensitivity number for each element using equation (4.17);

Step 4: Remove a number of elements which have the lowest sensitivity numbers;

Step 5: Repeat Steps 2 to 4 until one of the conditions in (4.16) can no longer be satisfied.

4.6 Example of Optimization with Multiple Displacement Constraints

A structure is to be designed to support three point loads, each of 10 kN, under the given boundary conditions shown in Figure 4.8. The dimensions for the design domain are L_x = 0.20 m, L_y = 0.10 m and thickness t = 0.005 m. The Young's modulus E = 207 GPa and the Poisson's ratio ν = 0.3 are assumed. The structure is assumed to be under plane stress conditions. The maximum initial vertical

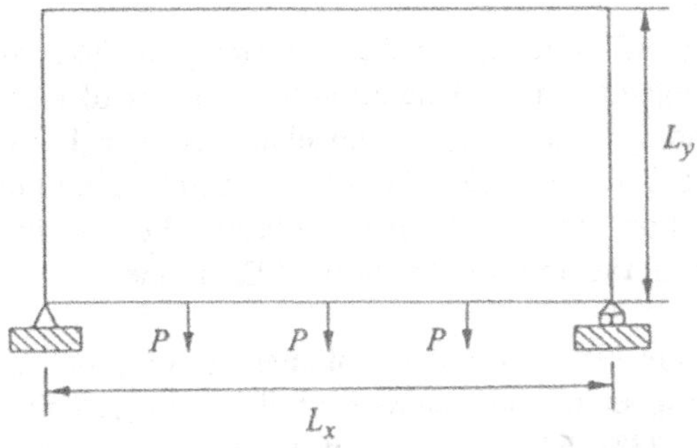

Fig. 4.8. Design domain for a structure under multiple displacement constraints.

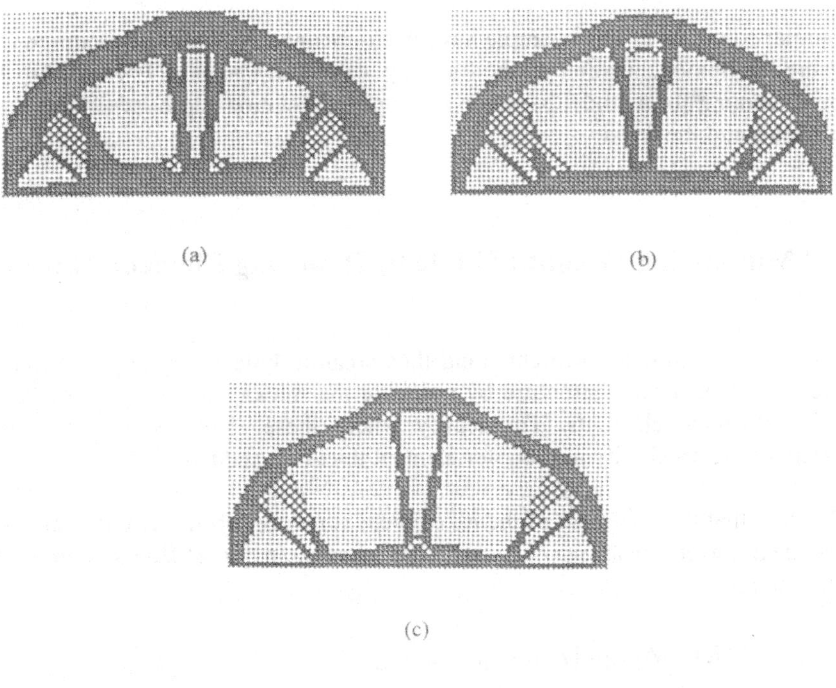

Fig. 4.9. Optimal designs for the structure under multiple displacement constraints for limits: (a) 0.30 mm; (b) 0.35 mm; (c) 0.45 mm.

displacement is 0.22 mm. Equal displacement limits are imposed simultaneously on all three loaded points in the vertical direction.

Due to symmetry only one half of the structure is analyzed using a mesh of 40 × 40 four node elements. At each iteration the number of elements removed is 16, which corresponds to 1% of the initial elements. The ESO results are given in Figure 4.9 for the different limits of 0.30, 0.35 and 0.45 mm on all loaded points in the vertical direction. The black areas represent the remaining elements and the small dots represent the nodes of the initial FEA model.

The ESO designs in Figure 4.9 have volumes of 50%, 39% and 30% of the initial volume. With respect to the volumes of the corresponding optimized uniform design, which are 73%, 63% and 50% of the initial volume, the ESO designs have resulted in volume reductions of 32%, 38% and 40%, respectively.

It is worth pointing out that by using the proposed weighting parameters in equation (4.18) the displacements in each optimum design at the three loaded points are all very close to the limit despite the fact that they are considerably different in the initial design.

Similar solutions to this example have been obtained by Díaz and Bendsøe (1992) using the homogenization method and by Yang and Chuang (1994) using linear programming for a model based on an empirical relationship between the Young's modulus and the density.

4.7 Minimizing Weight of Plate by Reducing Element Thickness

Rather than removing elements (and thus creating holes), we may also minimize the weight of a plate structure by reducing the thicknesses of some of the 'less efficient' plate elements. To identify where these elements are, we need to calculate the sensitivity number for element thickness reduction.

Suppose that the thickness of the *i*th element is reduced from the old thickness t to the next lower thickness $(t-\Delta t)$. The change in the stiffness matrix of the structure is

$$[\Delta K] = [\Delta K^i] = [K^i(t-\Delta t) - K^i(t)] \tag{4.19}$$

where $[K^i(t)]$ and $[K^i(t-\Delta t)]$ are the stiffness matrices of the *i*th element for the old thickness t and the new thickness $(t-\Delta t)$, respectively. If a constraint is imposed on the jth displacement component u_j, as given in equation (4.8) or (4.16), the effect of the element thickness reduction on u_j can be calculated, similar to equation (4.10), as

$$\alpha_{ij} = \Delta u_j = -\{u^{ij}\}^{\mathrm{T}}[\Delta K^i]\{u^i\}. \tag{4.20}$$

From the above equation, we define

$$\alpha_i = |\ -\{u^{ij}\}^{\mathrm{T}}[\Delta K^i]\{u^i\}|, \tag{4.21}$$

as the sensitivity number for the element thickness reduction. It is expected that reducing the thickness of the element with the smallest sensitivity number will result in the smallest change in the displacement.

The evolutionary procedure for least weight design of plate structures with variable thicknesses is given as follows:

Step 1: Discretize the structure using a fine mesh of finite elements;
Step 2: Solve the static equilibrium equation (4.1) for the given load $\{P\}$ and the virtual unit loads corresponding to all the displacement constraints;
Step 3: Calculate the sensitivity number for each element using equation (4.21);
Step 4: Reduce the thicknesses of a number of elements which have the lowest sensitivity numbers;
Step 5: Repeat Steps 2 to 4 until the displacement constraint can no longer be satisfied.

For problems with multiple displacement constraints, equation (4.17) may be employed to calculate a weighted average of the sensitivity numbers, which can then be used in Step 3.

The number of elements subjected to thickness reduction is determined by the step size Δt and the prescribed *material removal ratio* (*MRR*), defined as the amount (volume or weight) of material to be removed at each iteration over the total initial material. A typical value for *MRR* is 1%, which has been used to solve various problems by Chu *et al.* (1996c).

4.8 Example of Least Weight Design of Plate with Variable Thicknesses

Figure 4.10 shows a rectangular plate (4m × 2m) which is clamped at one long edge. There are three point loads, each of 20 kN, acting normal to the plate at the

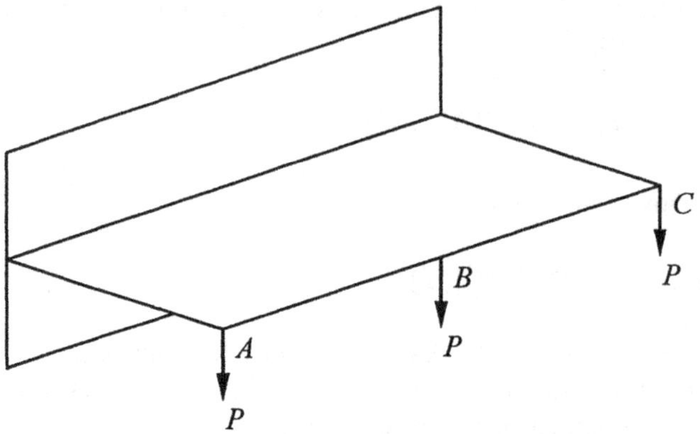

Fig. 4.10. A cantilever plate.

free edge at the two corners and in the middle. Multiple displacement constraints are imposed on the out-of-plane displacements at the loaded points. The Young's modulus $E = 30$ GPa and the Poisson's ratio $\nu = 0.2$ are assumed. The thickness of the plate t may vary from 0.1 to 0.2 m with equal intervals in between. For example, with five thicknesses, t may take the following discrete values: 0.1, 0.125, 0.15, 0.175 and 0.2 m.

Because of symmetry, only half of the cantilever is modelled using 400 four-noded plate elements. Initially, all elements are assigned the maximum thickness of 0.2 m. The initial maximum out-of-plane displacement is 2.22 mm. The limit imposed on the displacements at all three loaded points is 3.0 mm. For this displacement limit, the volume of the optimized uniform design is 90.5% of the initial volume. ESO solutions are obtained for the following cases:

(a) design for two thicknesses ($t = 0.1$ and 0.2 m, $\Delta t = 0.1$ m): With $MRR = 1\%$, eight elements are subjected thickness reduction at each iteration. The ESO design is shown in Figure 4.11(a), of which the volume is 78% of the initial volume.

(b) design for three thicknesses ($t = 0.1$, 0.15 and 0.2 m, $\Delta t = 0.05$ m): With $MRR = 1\%$, 16 elements are subjected thickness reduction at each iteration. The ESO design is shown in Figure 4.11(b), of which the volume is 79% of the initial volume.

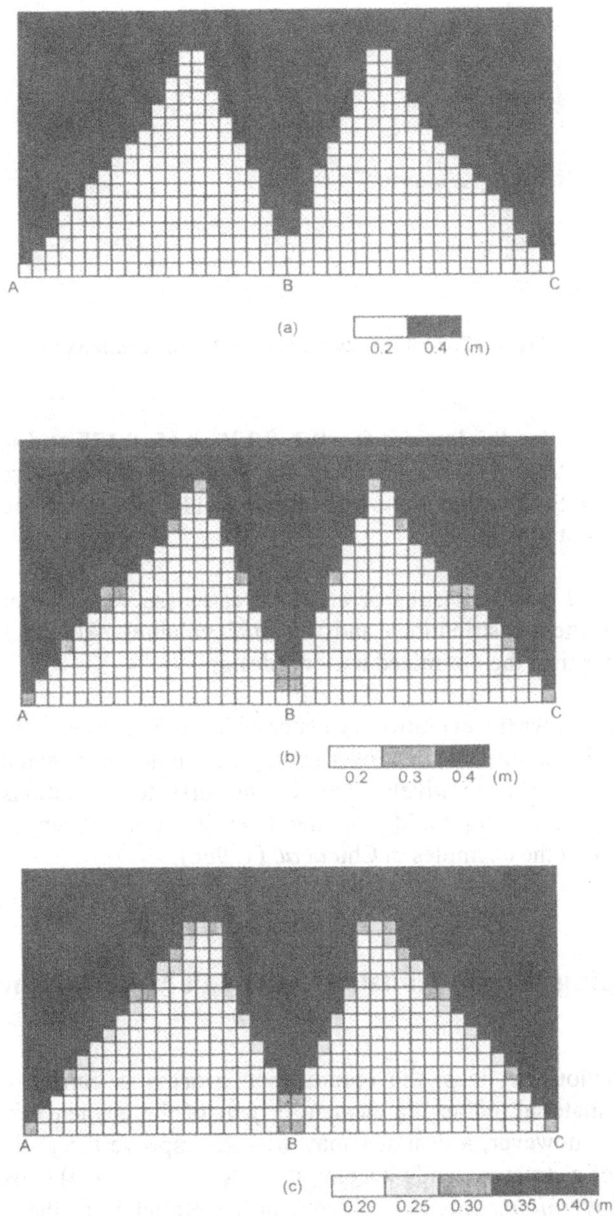

Fig. 4.11. Optimal designs for the cantilever plate: (a) for two thicknesses; (b) for three thicknesses; (c) for five thicknesses.

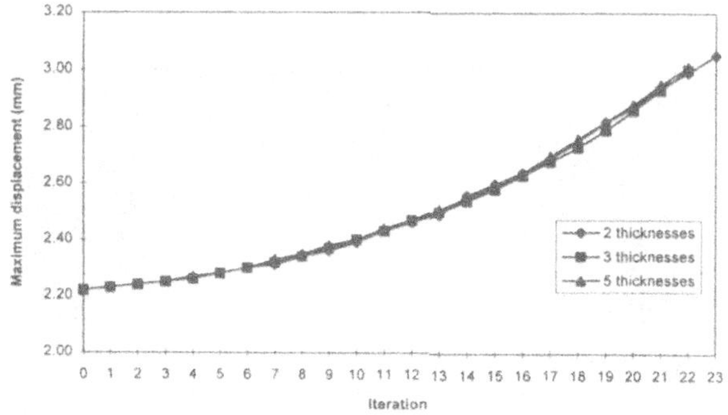

Fig. 4.12. Optimization histories for the cantilever plate.

(c) design for five thicknesses (t = 0.1, 0.125, 0.15, 0.175 and 0.2 m, Δt = 0.025 m): With *MRR* = 1%, 32 elements are subjected thickness reduction at each iteration. The ESO design is shown in Figure 4.11(c), of which the volume is again 79% of the initial volume.

It is observed that the results of the three cases are very similar in terms of the overall thickness distributions and the final volumes. All the ESO designs are much lighter than the optimized uniform design.

Figure 4.12 shows the evolutionary histories for all the cases, with the three curves for the maximum displacement overlapping each other. It is seen that the change in the displacement is relatively slow in the first few iterations and then the displacement increases rapidly in the later iterations. Such a trend has been observed in all the examples in Chu *at al.* (1996c).

4.9 Keeping Weight Constant Through Material Shifting

In the previous sections, the optimization process is carried out by gradually removing material (either the element or part of the element) from the structure. Sometimes, however, a designer may wish to improve the performance, e.g. the stiffness, of a structure while keeping its weight (or volume) constant. This can be achieved by *shifting* (instead of removing) material from the 'strongest' to the 'weakest' location. In the following, we describe this kind of material shifting technique for minimizing a specified displacement of a structure subject to constant weight. In particular we consider structures made of plate elements.

Assume that the objective is to minimize the absolute value of the *j*th displacement component, i.e. $|u_j|$. The effect of the thickness change of the *i*th element on u_j is given by equation (4.20). In the case of thickness increase we have

$$[\Delta K^i] = [\Delta K^i]^+ = [K^i(t+\Delta t)] - [K^i(t)], \tag{4.22}$$

and in the case of thickness reduction we have

$$[\Delta K^i] = [\Delta K^i]^- = [K^i(t-\Delta t)] - [K^i(t)]. \tag{4.23}$$

Therefore for the objective displacement $|u_j|$ two sensitivity numbers need to be calculated for each element, one for thickness increase

$$\alpha_i^+ = |\ -\{u^{ij}\}^{\mathrm{T}}[\Delta K^i]^+\{u^i\}|, \tag{4.24}$$

and the other for thickness reduction

$$\alpha_i^- = |\ -\{u^{ij}\}^{\mathrm{T}}[\Delta K^i]^-\{u^i\}|. \tag{4.25}$$

Generally if an element thickness is reduced, $|u_j|$ will increase; and if an element thickness is increased, $|u_j|$ will decrease. Therefore to minimize $|u_j|$ while keeping the weight constant it will be most effective to increase the thickness of the element with the highest α_i^+ and, at the same time, reduce the thickness of the element with the lowest α_i^-.

The evolutionary procedure for minimizing the objective displacement while keeping the weight constant for plate structures with variable thicknesses is given as follows :

Step 1: Discretize the structure using a fine mesh of finite elements;
Step 2: Solve the static equilibrium equation (4.1) for the given load $\{P\}$ and the virtual unit load corresponding to the objective displacement;
Step 3: Calculate the sensitivity numbers for α_i^+ and α_i^- for each element using equations (4.24) and (4.25);
Step 4: Reduce the thicknesses of a number of elements which have the lowest α_i^- and increase the thicknesses of the same number of elements with the highest α_i^+.
Step 5: Repeat Steps 2 to 4 until the objective displacement has converged to a steady state.

The number of elements subjected to thickness reduction (or increase) is determined by the step size Δt and the prescribed *material shifting ratio* (*MSR*), defined as the amount of material to be shifted at each iteration over the total initial material. Typical values for *MSR* are 0.5% and 1%, which have been used to solve various problems by Chu *et al.* (1996d).

In Step 5, a tolerance τ needs to be prescribed for the convergence check of the objective displacement. If the relative change in the objective displacement in two successive iterations is less that the given tolerance τ, i.e.

$$\left| \frac{u_j^{new} - u_j^{old}}{u_j^{new}} \right| < \tau , \tag{4.26}$$

then the steady state for the objective displacement is reached and the evolutionary procedure is terminated. A typical value for τ is 1%. To proceed the iterations beyond such a steady state will yield little improvement in the objective displacement.

4.10 Example of Optimal Design of Plate Subject to Constant Weight

A square plate (8m × 8m) is simply supported at four edges and is carrying a point load of 100 kN normal to the plate at its centre. The Young's modulus $E = 30$ GPa and the Poisson's ratio $\nu = 0.2$ are assumed. Initially the out-of-plane displacement at the centre is 4.22 mm. This displacement (the objective displacement) is to be minimized using the material shifting technique described in the preceding section.

Because of symmetry, only a quarter of the plate is analyzed using 20 × 20 four node plate elements. It is assumed that the thickness of the plate t may vary from 0.2 to 0.4 m with equal intervals in between. Initially all elements are assigned the average thickness of 0.3 m, which defines the weight of the plate. Using *MSR* = 1% and $\tau = 1\%$, ESO solutions are obtained for the following two cases:

(a) design for three thicknesses ($t = 0.2$, 0.3 and 0.4 m, $\Delta t = 0.1$ m): With *MSR* = 1%, there are 12 elements having thicknesses reduced and 12 increased at each iteration. After 14 iterations the solution converged and we obtain the optimal design of the plate which has kept the initial weight but reduced the objective displacement from 4.22 to 2.49 mm. The thickness distribution for a quarter of the optimized plate is shown in Figure 4.13(a).

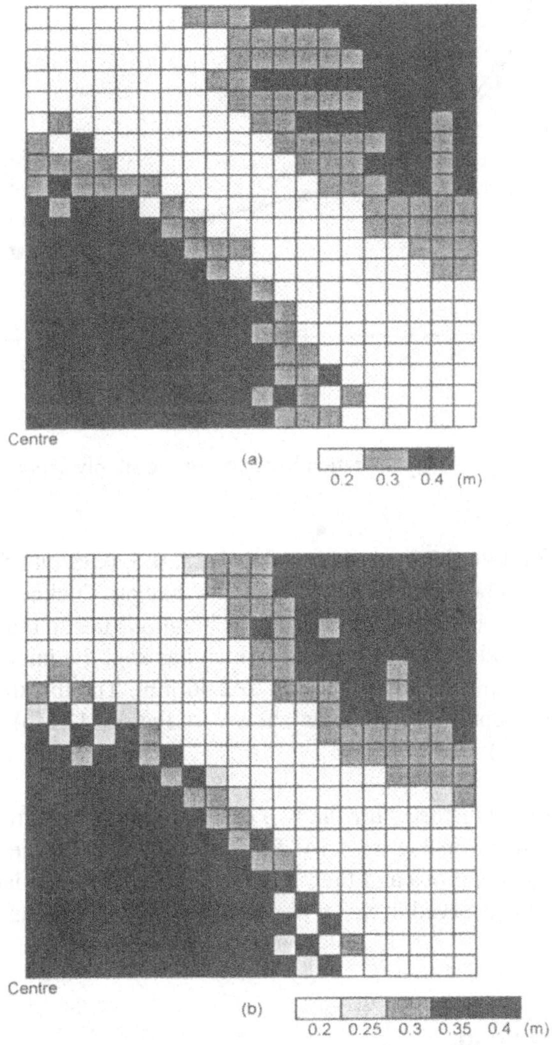

Fig. 4.13. Optimal designs for a quarter of the simply supported plate: (a) for three thicknesses; (b) for five thicknesses.

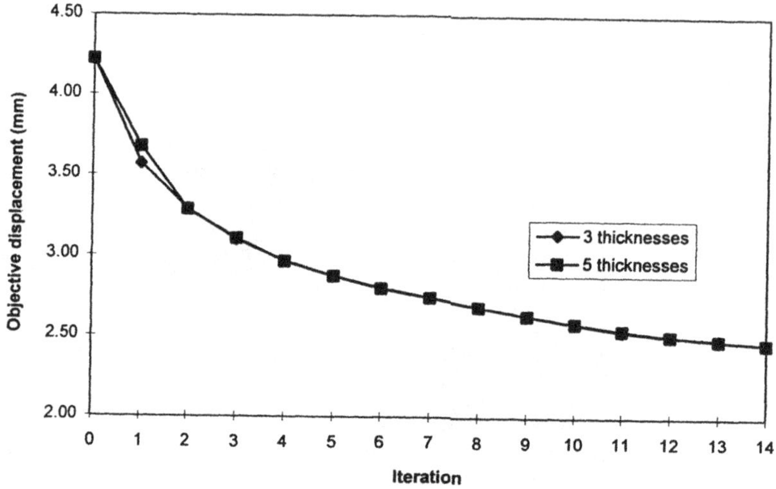

Fig. 4.14. Optimization histories for the simply supported plate.

(b) design for five thicknesses (t = 0.2, 0.25, 0.3, 0.35 and 0.4 m, Δt = 0.05 m): With MSR = 1%, there are 24 elements having thicknesses reduced and 24 increased at each iteration. After again 14 iterations the solution converged and we obtain the optimal design of the plate which has kept the initial weight but reduced the objective displacement from 4.22 to 2.46 mm. The thickness distribution for a quarter of the optimized plate is shown in Figure 4.13(b), which is obviously similar to 4.13(a).

The evolutionary histories for the two cases are almost identical except in the first iteration, as shown in Figure 4.14. The objective displacement is sharply reduced in the first few iterations and then it decreases slowly in the later iterations. This is a general trend observed in this type of optimization solutions.

4.11 Conclusion

This chapter has presented ESO procedures for various problems involving stiffness or displacement constraints, including multiple displacement constraints and a constraint on the difference of two displacements. Structural optimization can be achieved by removing elements, or reducing the thicknesses of elements, or shifting material from one element to another. A range of examples are presented for various problems and the results clearly demonstrated the effectiveness of the simple techniques described in this chapter.

4.12 References

Atrek, E. (1989) SHAPE: A program for shape optimization of continuum structures. In: C.A. Brebbia and S. Hernandez (eds.), *Computer Aided Optimum Design of Structures: Applications*, Computational Mechanics Publications, Southampton, 135-144.

Chu, D.N., Xie, Y.M., Hira, A. and Steven, G.P. (1996a) Evolutionary structural optimization for problems with stiffness constraints. *Finite Elements in Analysis and Design* **21**, 239-251.

Chu, D.N., Xie, Y.M., Hira, A. and Steven, G.P. (1996b) On various aspects of evolutionary structural optimization for problems with stiffness constraints. *Finite Elements in Analysis and Design* (to appear).

Chu, D.N., Xie, Y.M. and Steven, G.P. (1996c) An evolutionary structural optimization method for least weight design of plates with discrete variable thicknesses, *Comput. Struct.* (submitted).

Chu, D.N., Xie, Y.M. and Steven, G.P. (1996d) An evolutionary method for optimal design of plates with discrete variable thicknesses subject to constant weight, *Struct. Optim.* (submitted).

Díaz, A.R. and Bendsøe, M.P. (1992) Shape optimization of structures for multiple loading conditions using a homogenization method. *Struct. Optim.* **4**, 17-22.

Hemp, W.S. (1973) *Optimum Structures*, Clarendon Press, Oxford.

Jog, C., Haber, R.B. and Bendsøe, M.P. (1994) Variable-topology shape optimization with a constraint on the perimeter. *Proc. 20th ASME Design Automation Conf.*, September 11-14, 1994, Minneapolis, ASME publication DE-Vol. 69-2, 261-272.

Michell, A.G.M. (1904) The limits of economy of material in frame-structures. *Phil. Mag.* **8**, 589-597.

Olhoff, N., Bendsøe, M.P. and Rasmussen J. (1991) On CAD-integrated structural topology and design optimization. *Comput. Methods Appl. Mech. Eng.* **89**, 259-279

Zhou, M. and Rozvany, G.I.N. (1991) The COC algorithm, Part II: topological, geometrical and generalized shape optimization. *Comput. Methods Appl. Mech. Eng.* **89**, 309-336.

Tenek, L.H. and Hagiwara, I. (1994) Optimal rectangular plate and shallow shell topologies using thickness distribution or homogenization. *Comput. Methods Appl. Mech. Eng.* **115**, 111-124.

Venkayya, V.B., Knot, N.S. and Berke, L. (1973) Application of optimality criteria approaches to automated design of large practical structures. *Proc. AGARD Second Symposium on Structural Optimization*, April 2-6, 1973, Milan, pp. 3-1 to 3-18.

Yang, R.J. and Chuang, C.H. (1994) Optimal topology design using linear programming. *Comput. Struct.* **52**, 265-275.

Chapter 5 Frequency Optimization

The response of a structure to dynamic loading depends, to a large extent, on the first few natural frequencies of the structure. Excessive vibration occurs when the frequency of the dynamic loading is close to one of the natural frequencies of the structure. The optimum design of structures with frequency constraints is of great importance, particularly in the aeronautical and automotive industries. In order to avoid severe vibration, it is often necessary to shift the fundamental (first) frequency or several of the lower frequencies of the structure away from the frequency range of the dynamic loading. This chapter discusses the evolutionary procedures for various frequency optimization problems, which include maximizing or minimizing a chosen frequency, keeping a chosen frequency constant, maximizing the gap of two frequencies, as well as considerations of multiple frequency constraints. Much of the material presented in this chapter is based on the work of Xie and Steven (1996).

5.1 Introduction

There has been extensive research focused on structural optimization with frequency constraints in the past three decades. A survey by Grandhi (1993) covers most of the developments and applications in this area. It is noted that studies of frequency optimization have been mainly restricted to changing the size of beams or the thickness of plates etc. More recently attempts on the simultaneous shape and topology optimization of general structural dynamic systems have achieved some success, most notably the applications of the homogenization method (e.g. Díaz and Kikuchi 1992; Ma *et al.* 1993; Tenek and Hagiwara 1993; Ma *et al.* 1995).

We shall show in this chapter that the ESO concept of gradually removing material can be easily extended to and effectively used for frequency optimization. To date many frequency optimization problems have been solved using ESO and the results are in excellent agreement with solutions of other techniques such as the homogenization method. The great advantage of the ESO method is its simplicity compared with any other structural optimization method.

5.2 Sensitivity Number for Frequency

To identify the best locations for structural modifications, sensitivity derivatives are often needed. Usually the sensitivity analysis is very complicated, particularly for dynamic problems. However, we shall show in this section that a sensitivity number which indicates the effect of removing an element on a natural frequency can be easily calculated.

In finite element analysis the dynamic behaviour of the structure is represented by the following general eigenvalue problem

$$([K]-\omega_n^2[M])\{u_n\}=\{0\}, \tag{5.1}$$

where $[K]$ is the global stiffness matrix, $[M]$ is the global mass matrix, ω_n is the nth natural frequency and $\{u_n\}$ is the eigenvector corresponding to ω_n. The natural frequency ω_n and the corresponding eigenvector $\{u_n\}$ are related to each other by the Rayleigh quotient

$$\omega_n^2=\frac{k_n}{m_n}, \tag{5.2}$$

in which the modal stiffness k_n and the modal mass m_n are defined as

$$k_n=\{u_n\}^{\mathrm{T}}[K]\{u_n\}, \tag{5.3a}$$
$$m_n=\{u_n\}^{\mathrm{T}}[M]\{u_n\}. \tag{5.3b}$$

Suppose that we remove the ith element from the current structure. The change of the frequency due to such an element removal can be derived from equation (5.2) as

$$\Delta(\omega_n^2)=\frac{\Delta k_n}{m_n}-\frac{k_n\Delta m_n}{m_n^2}=\frac{1}{m_n}(\Delta k_n-\omega_n^2\Delta m_n). \tag{5.4}$$

To obtain the value of $\Delta(\omega_n^2)$ from the information of the previous eigenvalue solution, we assume that the eigenvector $\{u_n\}$ is approximately the same before and after the removal of that element. The assumption that the mode shape does not change significantly in between design cycles has been commonly used in frequency optimization (Grandhi 1993). We therefore have

$$\Delta k_n \approx \{u_n\}^{\mathrm{T}}[\Delta K]\{u_n\} = -\{u_n^i\}^{\mathrm{T}}[K^i]\{u_n^i\}, \tag{5.5a}$$

$$\Delta m_n \approx \{u_n\}^{\mathrm{T}}[\Delta M]\{u_n\} = -\{u_n^i\}^{\mathrm{T}}[M^i]\{u_n^i\}, \tag{5.5b}$$

in which $[K^i]$ and $[M^i]$ are the stiffness and mass matrices of the *i*th element, and $\{u_n^i\}$ is the eigenvector of the *i*th element. If, for example, a four node plane stress element is used $[K^i]$ and $[M^i]$ will be 8×8 matrices and $\{u_n^i\}$ will be a 8×1 vector.

Substituting approximations (5.5a) and (5.5b) into equation (5.4), we obtain the approximation of the change of the frequency due to the removal of the *i*th element, i.e.

$$\Delta(\omega_n^2) \approx \frac{1}{m_n}\{u_n^i\}^{\mathrm{T}}(\omega_n^2[M^i] - [K^i])\{u_n^i\}. \tag{5.6}$$

To decide which element should be removed from the structure so that the frequency will be shifted towards a desired value, we calculate for each element the sensitivity number for frequency as follows:

$$\alpha_n^i = \frac{1}{m_n}\{u_n^i\}^{\mathrm{T}}(\omega_n^2[M^i] - [K^i])\{u_n^i\}. \tag{5.7}$$

This sensitivity number is an indicator of the change in ω_n^2 as a result of the removal of the *i*th element.

In modal analysis it is a common practice that the eigenvector $\{u_n\}$ is normalized with respect to the mass matrix $[M]$, i.e. the eigenvector $\{u_n\}$ is scaled to make the modal mass m_n equal to unity. In this case, the formula for the sensitivity number α_n^i can be simplified by omitting m_n from equation (5.7). Even when the eigenvector $\{u_n\}$ is not normalized, the modal mass m_n, which is the same for every element, can still be omitted from equation (5.7) if we are concerned with only one frequency. However, when multiple frequencies are being considered, for example in the case of maximizing the gap of two frequencies, m_n has to be included in the calculation of the sensitivity number unless all the corresponding eigenvectors have been normalized with respect to $[M]$.

Fig. 5.1. The sensitivity number for frequency.

It is noted from equation (5.7) that the sensitivity number α_n^i can be easily calculated at the element level once the eigenvalue problem (5.1) is solved. The computational cost involved in calculating the sensitivity number is nominal compared to the cost of solving the eigenvalue problem.

Assume that the values of α_n^i are in the range of $[\alpha_{min}, \alpha_{max}]$, where α_{min} and α_{max} are the minimum and maximum values of α_n^i found over the whole design domain. α_{min} is less than zero and α_{max} is greater than zero as shown in Figure 5.1 (also see equation (5.8)). There are three regions in Figure 5.1 which are of importance to the frequency optimization. These regions are marked as A, B and C. For example, to increase the frequency it would be most effective, according to the definition of the sensitivity number, to remove those elements whose values of α_n^i fall in region C. Similarly, to reduce the frequency it would be most effective to remove elements whose values of α_n^i fall in region A. In other circumstances, the designer might be satisfied with the frequency ω_n of the current structure but wishes to reduce the structural weight while keeping ω_n constant (or changed as little as possible). This can be achieved by removing elements whose values of α_n^i are close to zero (i.e. in region B). From the definition of the Rayleigh quotient in equations (5.2) and (5.3), it can be proved that

$$\sum_{elem.} \alpha_n^i = 0 \tag{5.8}$$

namely, the summation of the sensitivity numbers over all the elements is equal to zero. When a computer code is written for the calculation of the sensitivity number α_n^i, equation (5.8) is very useful for checking the correctness of the code.

5.3 Evolutionary Procedures for Frequency Optimization

With the evolutionary structural optimization, the frequency of a structure can be shifted towards a desired value by gradually removing material from the structure. At each iteration, the eigenvalue problem (5.1) is solved and then a number of elements, for example 10 elements or one per cent of the total elements, are removed from the structure. This evolutionary process continues until an optimum design is found. In what follows we outline the evolutionary procedures for various types of frequency optimization.

5.3.1 Increase a Chosen Frequency ω_n

Step 1: Discretize the structure using a fine mesh of finite elements;
Step 2: Solve the eigenvalue problem (5.1);
Step 3: Calculate the sensitivity number α_n^i using equation (5.7) for each element;
Step 4: Remove a number of elements which have the highest α_n^i values;
Step 5: Repeat Steps 2 to 4 until an optimum is reached.

There are several ways to terminate the evolution in Step 5. The simplest way is by prescribing the maximum iteration number (say 100). A similar strategy is to prescribe the maximum amount of material (say 50%) that is allowed to be removed from the structure. The evolution can also be stopped when the frequency under consideration has reached the prescribed target value.

5.3.2 Reduce a Chosen Frequency ω_n

The solution procedure for reducing a chosen frequency is similar to that in Section 5.3.1 The only difference is in Step 4, which should be re-written for the present case as follows:
Step 4: Remove a number of elements which have the *lowest* α_n^i values;

5.3.3 Keep a Chosen Frequency ω_n Constant

When material is removed from a structure, the frequencies of the structure will almost inevitably be changed to some extent. What we try to do here is to make as little change as possible in a chosen frequency while the weight of the structure is being reduced. Again the solution procedure is the same as that in Section 5.3.1, except for Step 4 which should be restated as:
Step 4: Remove a number of elements which have the smallest $|\alpha_n^i|$ values.

5.3.4 Increase the Gap Between Two Frequencies

In some cases, for example in the design of certain flexible space vehicles, it is necessary to separate closely spaced frequencies so that a frequency based control system can work effectively. Suppose that we wish to increase the gap between two natural frequencies ω_p and ω_q. Following the derivation of the sensitivity number in Section 5.2, it is easy to obtain a similar sensitivity number for $(\omega_p^2 - \omega_q^2)$ as

$$\alpha_{pq}^i \approx \Delta(\omega_p^2 - \omega_q^2) \approx \alpha_p^i - \alpha_q^i, \qquad \text{where } p > q. \tag{5.9}$$

α_{pq}^i indicates the effect of removing the *i*th element on the change in the distance between ω_p^2 and ω_q^2. Once the sensitivity numbers α_p^i and α_q^i for individual modes are calculated using equation (5.7), the new sensitivity number α_{pq}^i can be immediately obtained from equation (5.9). The evolutionary procedure for increasing the gap between two frequencies is again the same as that in Section 5.3.1, except that Step 4 needs to be restated as:

Step 4: Remove a number of elements which have the highest α_{pq}^i values.

In practice the designer would be only interested in increasing the gap between two neighbouring frequencies, although the above procedure allows him to maximize the distance between two arbitrarily chosen frequencies.

5.3.5 Optimization with Multiple Frequency Constraints

The optimization of a structure with multiple frequency constraints can be very complicated. However, it is possible to extend the simple evolutionary procedure in Section 5.3.1 to optimization with considerations of multiple frequencies.

Suppose that we wish to increase the first three frequencies. After solving the eigenvalue problem (5.1), the sensitivity numbers α_1^i, α_2^i and α_3^i for the first three modes are calculated for each element using equation (5.7). In order to have the first three frequencies increased, only elements whose values of α_1^i, α_2^i and α_3^i are all greater than zero should be removed. If we are most concerned with the first frequency, Step 4 in Section 5.3.1 may be restated as:

Step 4: Remove elements whose values of α_1^i are the highest among those elements which have $\alpha_1^i > 0$, $\alpha_2^i > 0$, and $\alpha_3^i > 0$.

5.4 Examples

The examples presented in this section are all two-dimensional plane stress problems. The element used is the four node linear element. However, the ESO method described in this chapter is not restricted to any particular element and it can be equally applied to three-dimensional problems.

5.4.1 A Short Beam

Figure 5.2 shows a short beam of dimensions 5m × 1m. The beam is clamped on both sides. Young's modulus E = 200 GPa, Poisson's ratio = 0.3, thickness t = 0.1 m and density $\rho = 7000\ \mathrm{kg/m^3}$ are assumed. The design domain is divided equally into 100 × 20 square elements. Four elements are removed at the end of each finite element analysis.

As the first test we try to increase the fundamental frequency of this short beam. Figure 5.3 shows the evolutionary history of the first four frequencies of the beam during the first 50 iterations while the first frequency is being increased. This is achieved by removing, at each iteration, four elements which have the highest α_1^i values among all the remaining elements in the current structure. A new design of the beam corresponding to the 50th iteration is given in Figure 5.4. At this stage 10% of the total material has been eliminated and the first frequency has been increased from 178.4 to 199.3 Hz, an increase of 11.7%.

Similarly the second frequency of the beam can be increased by removing elements which have the highest α_2^i values. The evolutionary history of the first four frequencies for this test is given in Figure 5.5. After 50 iterations a new design is obtained as shown in Figure 5.6. The material reduction is again 10% and the second frequency is increased by 9.5% from 409.9 to 448.7 Hz.

It is interesting to note that the new designs of the short beam shown in Figures 5.4 and 5.6 are very similar to what Tenek and Hagiwara (1993) obtained using the homogenization method.

Next we consider increasing the gap between the first and the second frequencies of the short beam. To increase the gap between ω_1 and ω_2 we remove, at each iteration, four elements which have the highest $(\alpha_2^i - \alpha_1^i)$ values. Figure 5.7(a) shows the evolutionary history of the first four frequencies. The increase in

Fig. 5.2. A short beam under plane stress conditions.

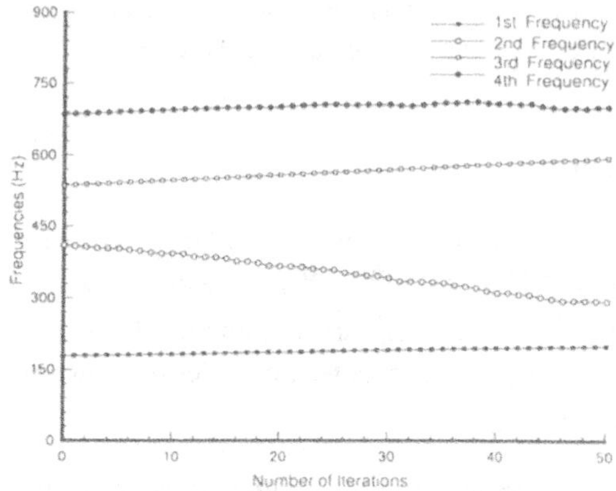

Fig. 5.3. Evolutionary history of the first four frequencies of the short beam: increasing the first frequency.

Fig. 5.4. A new design of the short beam with increased first frequency. Material removed: 10%; increase in ω_1 : 11.7%.

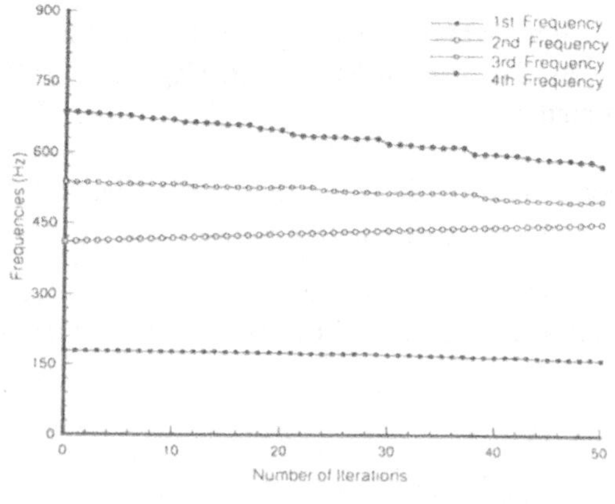

Fig. 5.5. Evolutionary history of the first four frequencies of the short beam: increasing the second frequency.

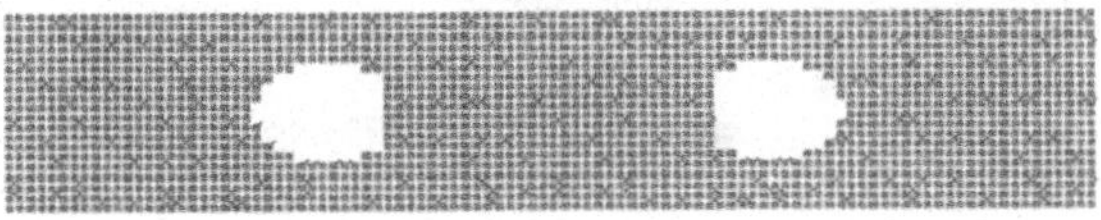

Fig. 5.6. A new design of the short beam with increased second frequency. Material removed: 10%; increase in ω_2: 9.5%.

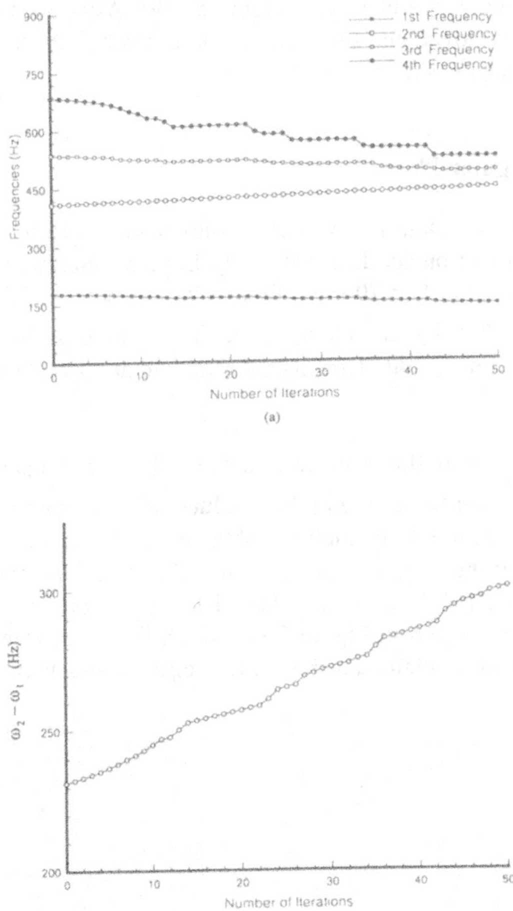

Fig. 5.7. Evolutionary histories for increasing the gap between ω_1 and ω_2 of the short beam: (a) the first four frequencies; (b) the gap between ω_1 and ω_2.

Fig. 5.8. A new design of the short beam with increased $(\omega_2 - \omega_1)$. Material removed: 10%; increase in$(\omega_2 - \omega_1)$: 30.1%.

$(\omega_2 - \omega_1)$ is shown more clearly in Figure 5.7(b). After 50 iterations, 10% of the total material is removed and $(\omega_2 - \omega_1)$ is increased by 30.1% from 231.5 to 301.2 Hz. The new design is shown in Figure 5.8.

5.4.2 A Rectangular Plate

Figure 5.9 shows an aluminium plate of dimensions 0.15m × 0.1m. The plate is fixed at two corners on its diagonal. Only in-plane vibration is considered here. The Young's modulus E = 70 GPa, Poisson's ratio $\nu = 0.3$, thickness t = 0.01 m and density $\rho = 2700\ \text{kg/m}^3$ are assumed. The domain is divided equally into 45 × 30 square elements. Eight elements are removed at the end of each finite element analysis.

Using the procedure outlined in Section 5.3.1 the first frequency is increased by removing the elements with the highest values of α_1^i. As a result the evolutionary history of the first three frequencies is obtained as shown in Figure 5.10. After 85 iterations, 50% of the total material is removed and the first frequency is increased by 39.0% from 2441.5 to 3392.8 Hz. The corresponding new design of the rectangular plate is given in Figure 5.11, which is almost identical to what Tenek and Hagiwara (1993) obtained using the homogenization method.

Fig. 5.9. A rectangular plate under plane stress conditions.

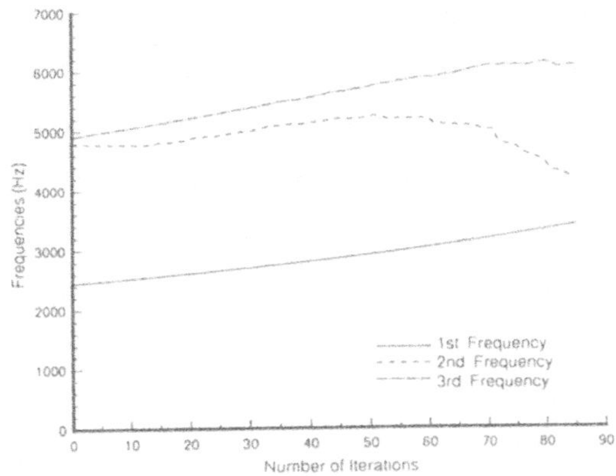

Fig. 5.10. Evolutionary history of the first three frequencies of the rectangular plate: increasing the first frequency.

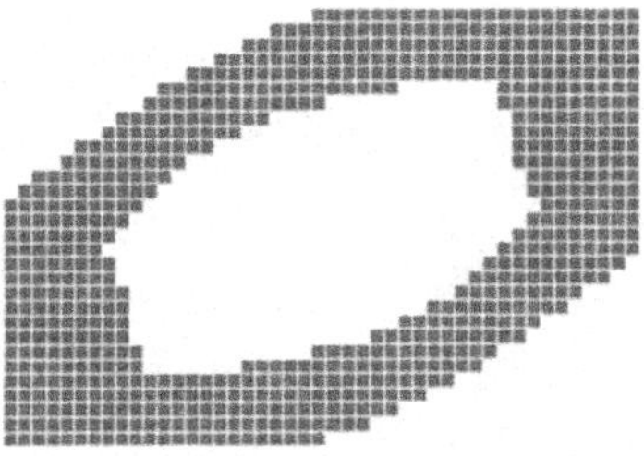

Fig. 5.11. A new design of the rectangular plate with increased first frequency. Material removed: 50%; increase in ω_1 : 39.0%.

In the previous test, only the first frequency has been considered for optimization. There is no control over what is happening to other frequencies during the optimization process. It is clear from Figure 5.10 that while the first frequency has been increasing monotonically, the second frequency starts to drop after about 50 iterations and its value at the 85th iteration is even smaller than its initial value. Sometimes, such behaviour is undesirable. This can be avoided, or at least delayed, by using the procedure for optimization with multiple frequency constraints as outlined in Section 5.3.5. The results are shown in Figures 5.12 and 5.13. It is seen from Figure 5.12 that all the first three frequencies have been increasing for the first 85 iterations, except that ω_2 has slight oscillations towards the end. The new design in Figure 5.13 has an extra bar in the middle compared to the solution in

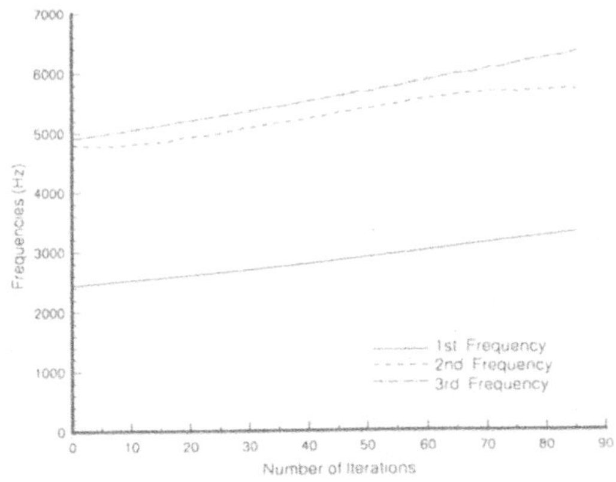

Fig. 5.12. Evolutionary history of the first three frequencies of the rectangular plate: increasing the first three frequencies.

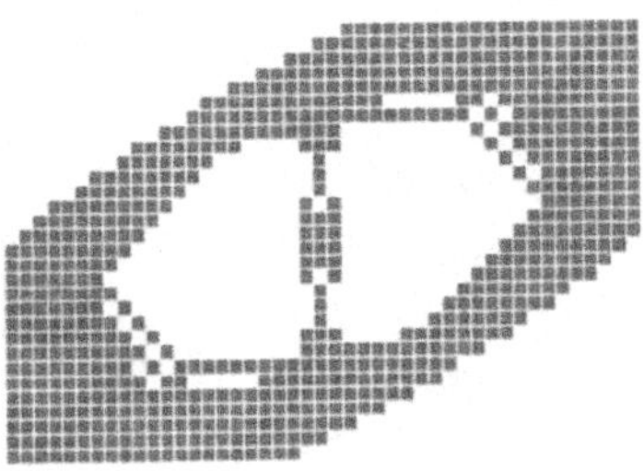

Fig. 5.13. A new design of the rectangular plate with increased first three frequencies. Material removed: 50%; increase in ω_1 : 35.5%; increase in ω_2 : 19.1%; increase in ω_3 : 29.3%.

Figure 5.11. Again 50% of the total material is removed, the first three frequencies are increased by 35.5%, 19.1% and 29.3%, respectively.

5.4.3 A Square Plate

A square plate of dimensions 1m × 1m is clamped on three sides as shown in Figure 5.14. The Young's modulus $E = 200$ GPa, Poisson's ratio $\nu = 0.3$, thickness $t = 0.01$ m and density $\rho = 7000 \text{ kg/m}^3$ are assumed. The design domain is

Fig. 5.14. A square plate under plane stress conditions.

Fig. 5.15. A new design of the square plate with increased first frequency. Material removed: 8%; increase in ω_1 : 14.3%.

divided equally into 50 × 50 square elements. Eight elements are removed at the end of each finite element analysis.

First we try to increase the fundamental frequency. Figure 5.15 shows the new shape of the square plate after 25 iterations when 8% of the total material is removed. The first frequency is increased by 14.3% from 2024.1 to 2313.3 Hz.

Figure 5.16 is the solution for reducing the first frequency at the 25th iteration. The material reduction is again 8%, but the first frequency is reduced considerably from 2024.1 to 162.7 Hz, a decrease of 92.0%. This is due to the slim "neck"

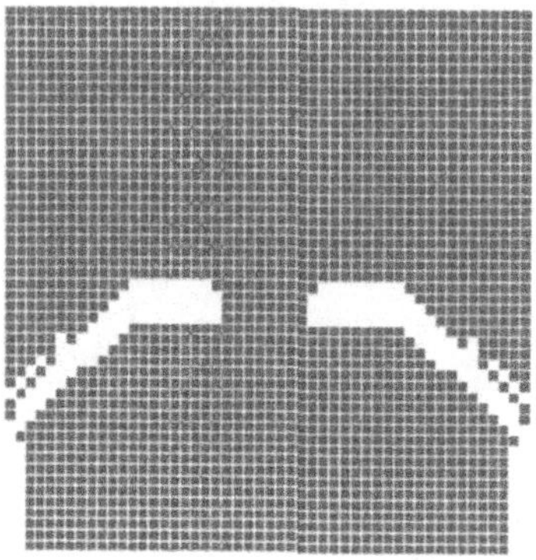

Fig. 5.16. A new design of the square plate with reduced first frequency. Material removed: 8%; reduction in ω_1 : 92.0%.

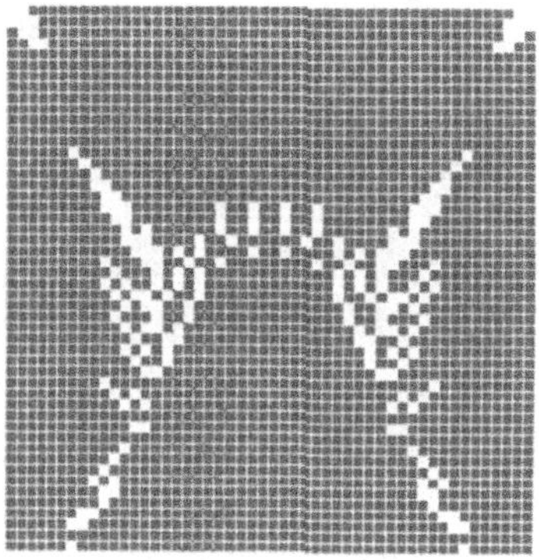

Fig. 5.17. A new design of the square plate with minimum change in the first frequency. Material removed: 8%; change in ω_1 : - 6.7%.

connecting the top and bottom parts, making the structure much more flexible.

The next attempt is made to keep the change in the first frequency as small as possible using the procedure outlined in Section 5.3.3. Figure 5.17 shows the result after 25 iterations. The change (reduction) in the first frequency is 6.7% after 8% of the total material has been removed.

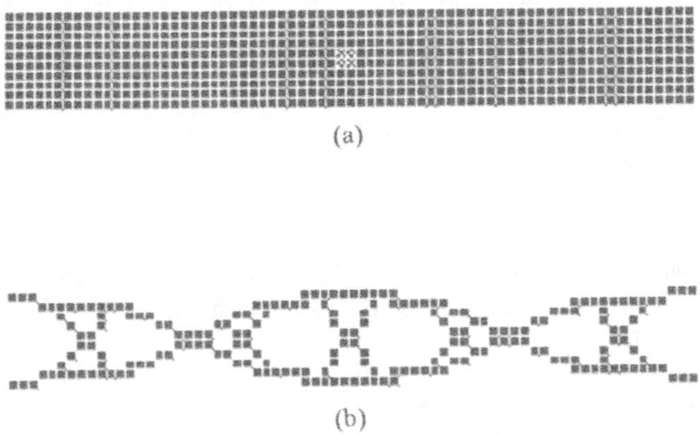

Fig. 5.18. Optimization of a short beam with a lumped mass. (a) design domain; (b) optimum design.

5.4.4 A Short Beam with Lumped Mass

Similar to the first example, a short beam is clamped at both ends as shown in Figure 5.18(a). In this example, however, there exists a lumped mass at the centre of the beam. Details of this example can be found in Zhao *et al.* (1995), where a few extra criteria with regard to element removal are added to the present evolutionary procedure for frequency optimization. These criteria include the positive definite structure criterion, the smooth change criterion and the average value criterion (Zhao *et al.* 1995, 1996). The optimum design obtained by Zhao *et al.* (1995) for this short beam is shown in Figure 5.18(b). A similar solution has been obtained by Ma *et al.* (1995) using the homogenization method.

5.5 Conclusion

In this chapter we have demonstrated that the ESO method can effectively solve a wide range of frequency optimization problems. The location for element removal is determined by the sensitivity number of each element. The sensitivity number can be easily calculated using the information available from the finite element solution. By gradually removing material from the initial design, we can increase or reduce the frequency of a structure or we can keep the frequency "constant".

The number of elements to be removed at each step is worth further investigation. The numbers we have used for the examples in this chapter correspond to removing at each iteration from 0.2 to 0.6% of the total initial material.

5.6 References

Díaz, A.R. and Kikuchi, N. (1992) Solutions to shape and topology eigenvalue optimization problems using a homogenization method. *Int. J. Numer. Methods Eng.* **35**, 1487-1502.

Grandhi, R.V. (1993) Structural optimization with frequency constraints - a review. *AIAA J.* **31**(12), 2296-2303.

Ma, Z.-D., Kikuchi, N., Cheng, H.-C. and Hagiwara, I. (1993) Topology and shape optimization methods for structural dynamic problems. In: P. Pedersen (ed.), *Optimal Design with Advanced Materials,* Elsevier Science Publishers, Amsterdam, 247-261.

Ma, Z.-D., Kikuchi, N. and Cheng, H.-C. (1995) Topological design for vibrating structures. *Comput. Methods Appl. Mech. Eng.* **121**, 259-280.

Tenek, L.H. and Hagiwara, I. (1993) Static and vibrational shape and topology optimization using homogenization and mathematical programming. *Comput. Methods Appl. Mech. Eng.* **109**, 143-154.

Xie, Y.M. and Steven, G.P. (1996) Evolutionary structural optimization for dynamic problems. *Comput. Struct.* **58**(6), 1067-1073.

Zhao, C., Steven, G.P. and Xie, Y.M. (1995) Evolutionary natural frequency optimization of two dimensional structures with nonstructural lumped masses. Research Report FEARC-9513, Finite Element Analysis Research Centre, University of Sydney, Australia.

Zhao, C., Steven, G.P. and Xie, Y.M. (1996) Evolutionary natural frequency optimization of thin plate bending vibration problems. *Struct. Optim.* **11**, 244-251.

Chapter 6 Optimization Against Buckling

With the advent of high strength materials, many structural elements are becoming thinner and more slender which makes them more susceptible to buckling. Optimum structural design against buckling may be achieved by shifting material from the strongest locations to the weakest so that the final design may have a much higher buckling load than that of the initial design of the same weight. This chapter describes the evolutionary procedure for the optimization of columns, frames and plates against buckling. Much of the material presented in this chapter derives from the work by Manickarajah *et al.* (1995, 1996).

6.1 Introduction

For a slender member, the load carrying capacity is often determined by its buckling load. In spite of extensive research in structural optimization, only a small number of works have dealt with buckling optimization. First, Keller (1960) presents the shape of the strongest column with simply supported ends. Later, Tadjbakhsh and Keller (1962) find the strongest column for several other boundary conditions. Simitses *et al.* (1973) appear to be the first to report the use of finite element analysis and iterative procedure to optimize the shape of columns. This method is extended to optimization of frames by Szyszkowski and Watson (1988) and Szyszkowski *et al.* (1989).

The work on buckling optimization of plate structures is more limited. Spillers and Levy (1990) extend the classic solution of Keller (1960) to a simply supported plate. Unfortunately, the assumptions used by Spillers and Levy are wrong and their design is in fact much worse than the uniform thickness plate of the same weight, as pointed out by Pandey and Sherbourne (1992). Then Pandey and Sherbourne (1992) present a new thickness distribution for the simply supported square plate which results in an increase in the buckling load of the plate by 28% compared to the uniform thickness plate of the same weight.

This chapter extends the ESO method to buckling optimization. Optimization against buckling may be obtained by finding the minimum weight design of a structure that satisfies the buckling load constraint. Alternatively it may be achieved by maximizing the buckling load of the structure while keeping its weight constant. The former approach may start with an over-designed structure with excess weight followed by gradually removing material from the structure. The latter approach involves gradually shifting material from the strongest locations to the weakest without adding or losing any weight. For the convenience of comparing the efficiency of different designs, the latter approach is adopted in this chapter. Note that in Chapter 4 both of the two approaches have been used.

It will be shown that the optimization of columns and frames against buckling can be easily achieved using the evolutionary procedure. Applications of the same evolutionary procedure to plates are also discussed.

6.2 Sensitivity Number for Buckling Load

The linear buckling behaviour of a structure is governed by the following general eigenvalue problem:

$$([K]+\lambda_n[K_g])\{u_n\}=\{0\}, \tag{6.1}$$

where $[K]$ is the global stiffness matrix, $[K_g]$ is the global *stress matrix* or *geometric stiffness matrix*, λ_n is the nth eigenvalue and $\{u_n\}$ is the corresponding eigenvector. The eigenvalues in equation (6.1) are those which scale the applied loading in the finite element model to give buckling. The most critical buckling load is of course the lowest one, which is equal to the first eigenvalue λ_1 multiplied by the applied loading in the finite element model. In this chapter λ_1 is referred to as the *buckling load factor*.

Multiplying equation (6.1) by the transpose of the eigenvector $\{u_n\}$ produces the following Rayleigh quotient for the eigenvalue:

$$\lambda_n=-\frac{\{u_n\}^{\mathrm{T}}[K]\{u_n\}}{\{u_n\}^{\mathrm{T}}[K_g]\{u_n\}}. \tag{6.2}$$

Consider a small change in the cross-sectional area of the ith element. Assuming that the eigenvector is approximately the same before and after such a small change, we obtain

$$\Delta\lambda_n = -\frac{\{u_n\}^{\mathrm{T}}([\Delta K]+\lambda_n[\Delta K_g])\{u_n\}}{\{u_n\}^{\mathrm{T}}[K_g]\{u_n\}}. \tag{6.3}$$

The change in the global stiffness matrix $[K]$ is equal to the change in the element stiffness matrix of the ith element, which can be easily calculated. However, since $[K_g]$ depends on the current stress distribution in the structure and the cross-sectional change in the ith element affects the stress in its surrounding elements, one cannot assume that $[\Delta K_g]$ is equal to the change in the element stress matrix of the ith element only. The calculation of $[\Delta K_g]$ is generally much involved. Fortunately $[\Delta K_g]$ is equal to zero if the axial or membrane stress resultant remains constant before and after the cross-sectional change in the elements. Such is the situation of all statically determinate structures. For a statically determinate frame, the cross-sectional changes do not affect the axial forces in the members. For a statically indeterminate structure, $[\Delta K_g]$ is only negligible if the cross-sectional modifications at each iteration are so small that they do not cause significant changes in the axial or membrane stress resultants. When $[\Delta K_g]$ is ignored, equation (6.3) is reduced to

$$\Delta\lambda_n = -\frac{\{u_n\}^{\mathrm{T}}[\Delta K]\{u_n\}}{\{u_n\}^{\mathrm{T}}[K_g]\{u_n\}} = -\frac{\{u_n^i\}^{\mathrm{T}}[\Delta K^i]\{u_n^i\}}{\{u_n\}^{\mathrm{T}}[K_g]\{u_n\}}, \tag{6.4}$$

where $\{u_n^i\}$ is the eigenvector of the ith element, $[K^i]$ is the stiffness matrix of the ith element. If the eigenvector is normalized with respect to $[K_g]$ so that the denominator in equation (6.4) equals unity, then equation (6.4) is further simplified to

$$\Delta\lambda_n = -\{u_n^i\}^{\mathrm{T}}[\Delta K^i]\{u_n^i\}. \tag{6.5}$$

The objective of buckling optimization is to raise the fundamental eigenvalue λ_1 so that the buckling load may be maximized. From the above equation, we define the sensitivity number for buckling load as follows:

$$\alpha_i = -\{u_1^i\}^{\mathrm{T}}[\Delta K^i]\{u_1^i\}. \tag{6.6}$$

The calculation of the above sensitivity number only involves small matrices of individual elements. This sensitivity number is used in this chapter as a measure of the effect of changing the cross-sectional area of the *i*th element on the buckling load factor. If the structure is statically determinate and the cross-sectional change at each iteration is small, the sensitivity number gives a very accurate estimation of the change in the buckling load factor. Even for a statically indeterminate frame this sensitivity is reasonably accurate if the cross-sectional variations in the frame results in only slight changes in the axial stress resultants. For a plate structure, this sensitivity number only works for the case of gradually varying the thicknesses of plate elements. If an element is removed from a plate or a frame, because of the significant changes in the membrane or axial stress resultants in its surrounding elements, equations (6.4) and (6.5) will be invalid and the sensitivity number defined in equation (6.6) will be incorrect too. Therefore, equation (6.6) should not be used for buckling optimization involving element removal. In this chapter the buckling optimization is carried out by changing the cross-sectional areas only.

In the case of an increase in the cross-sectional area A we have

$$[\Delta K^i] = [\Delta K^i]^+ = [K^i(A+\Delta A)] - [K^i(A)], \tag{6.7}$$

and in the case of a reduction in the cross-sectional area

$$[\Delta K^i] = [\Delta K^i]^- = [K^i(A-\Delta A)] - [K^i(A)]. \tag{6.8}$$

Hence to estimate the effect of cross-sectional changes on the buckling load factor two sensitivity numbers need to be calculated for each element, one for area increase

$$\alpha_i^+ = -\{u_1^i\}^{\mathrm{T}}[\Delta K^i]^+\{u_1^i\}, \tag{6.9}$$

and the other for area reduction

$$\alpha_i^- = -\{u_1^i\}^{\mathrm{T}}[\Delta K^i]^-\{u_1^i\}. \tag{6.10}$$

From the definition of the sensitivity number it is clear that to raise the buckling load factor it will be most effective to increase the cross-sectional areas of elements with the highest α_i^+ values and reduce those with the highest α_i^- values.

6.3 Bimodal and Multimodal Problems

During the process of buckling optimization, it is often observed that while the first eigenvalue is increasing the subsequent eigenvalues are decreasing and gradually the first two or more eigenvalues converge to each other, although the corresponding eigenvectors may remain totally different. When this happens, the structure is said to be bimodal or multimodal. Some symmetrical structures are intrinsically multimodal from the outset. The multimodal behaviour necessitates a mathematical reformulation of the single modal problem since the repeated eigenvalues are not continuously differentiable. Extensive research has been carried out to deal with the bimodal and multimodal problems (e.g. Olhoff and Rasmussen 1977; Haug 1982; Szyszkowski *et al.* 1989; Rodrigues *et al.* 1995).

With evolutionary structural optimization, the sensitivity number is the crucial factor. In equation (6.6) only the first eigenvector is considered. When the first eigenvalue becomes close to the subsequent eigenvalues, there will be interference between the first and the subsequent buckling modes. Therefore the effect on the first eigenvalue of all the participating eigenvectors needs to be included. If, for example, the distance between λ_1 and λ_2 is within a certain limit, say 5%, and the distance between λ_3 and λ_1 is greater than that limit, we may assume that the structure has now become bimodal. When λ_1 and λ_2 become close, the first two buckling modes may swap with each other as a result of structural modifications during the iterations. There is no point in trying to increase λ_1 only to see λ_2 to drop its value below previous λ_1 in the next step. To effectively increase the buckling load factor in these circumstances, we have to increase both λ_1 and λ_2. The simplest strategy for achieving this is to increase the average of λ_1 and λ_2. Therefore we redefine the sensitivity number for the bimodal case as

$$\alpha_i = \Delta\left(\frac{\lambda_1 + \lambda_2}{2}\right) = -\frac{1}{2}(\{u_1^i\}^{\mathrm{T}}[\Delta K^i]\{u_1^i\} + \{u_2^i\}^{\mathrm{T}}[\Delta K^i]\{u_2^i\}). \tag{6.11}$$

Similarly for the multimodal case, when the distance between the *m*th eigenvalue λ_m and the first eigenvalue λ_1 is within a certain limit, say 5%, and the distance between λ_{m+1} and λ_1 is greater than that limit, the sensitivity number is redefined as

$$\alpha_i = \Delta\left(\frac{1}{m}\sum_{j=1}^{m}\lambda_j\right) = -\frac{1}{m}\sum_{j=1}^{m}(\{u_j^i\}^{\mathrm{T}}[\Delta K^i]\{u_j^i\}). \tag{6.12}$$

These sensitivity numbers in equations (6.11) and (6.12) for bimodal and multimodal problems are extremely simple. They are the average values of the individual sensitivity numbers of all participating modes. This same concept can be applied to the frequency optimization problems when repeated frequencies occur.

It is worth emphasizing that to use equations (6.6) to (6.12) for the calculation of the sensitivity numbers all the relevant eigenvectors have to be first normalized with respect to $[K_g]$. For each element two sensitivity numbers, α_i^+ and α_i^-, need to be calculated. α_i^+ is obtained by substituting $[\Delta K^i]^+$ of equation (6.7) into (6.11) or (6.12) and α_i^- is obtained by substituting $[\Delta K^i]^-$ of equation (6.8) into (6.11) or (6.12).

6.4 Evolutionary Procedure for Buckling Optimization

The evolutionary procedure for buckling optimization of columns, frames and plates are outlined as follows:

Step 1: Discretize the structure using a fine mesh of finite elements;
Step 2: Solve the eigenvalue problem (6.1);
Step 3: Calculate the sensitivity numbers α_i^+ and α_i^- for each element;
Step 4: Increase the cross-sectional area of elements which have the highest α_i^+ values and decrease the cross-sectional area of the same number of elements which have the highest α_i^- values;
Step 5: Repeat Steps 2 to 4 until the buckling load factor cannot be increased any further.

During the evolution, the cross-sectional area is allowed to vary in small steps in a prescribed manner. For example, for beam elements with rectangular cross-sections, either the breadth or the depth may be changed; and for plate elements, the thickness may be changed. For such simple variations, the change in the element stiffness matrix, $[\Delta K^i]$, can be easily calculated. All the other information required for the calculation of sensitivity numbers is readily available from the finite element solution obtained in Step 2.

In the above evolutionary procedure, the number of elements subjected to cross-sectional changes and the step size of the change at each iteration have to be prescribed. There have been some discussions on these parameters by Chu *et al.* (1996) for stiffness optimization. The influence of these parameters on the final solution needs to be investigated for buckling optimization.

6.5 Example of Column Optimization

We consider a slender column of variable, but geometrically similar, cross-sections where the area $A(x)$ and the area moment of inertia $I(x)$ at section x are correlated by $I(x) = cA(x)^p$, with the constant c given by the cross-sectional geometry and p is equal to 1, 2 and 3. It is assumed that the column is simply supported at both ends and is subjected to a compressive axial load. The column is divided into 100 linear elements of equal length. Initially the column is of a uniform cross-section. The following three cases are considered:

Case 1: $p = 1$
The simplest cross-sectional shape which may satisfy the relationship $I(x) = cA(x)$ is a rectangle with constant depth d and variable width b so that $I(x) = b(x)\, d^3 / 12 = A(x)d^2 / 12$. The dimensions of the column are as follows: length $L = 1$ m, $d = 50$ mm, initial value for $b = 65$ mm. During the optimization, b is allowed to vary in steps of 5 mm to the maximum of 125 mm and to the minimum of 5 mm. Hence 25 different cross-sectional areas are allowed in the design.

Case 2: $p = 2$
For a circular cross-section, $I(x) = \pi r(x)^4 / 4 = A(x)^2 / 4\pi$, where $r(x)$ is the radius at section x. The dimensions of the column are as follows: length $L = 1$ m, initial value for $r^2 = 40$ mm^2. During the optimization r^2 is allowed to vary in steps of 1 mm^2 to the maximum of 60 mm^2 and to the minimum of 10 mm^2. Hence 51 different cross-sectional areas are allowed in the design.

Case 3: $p = 3$
For the rectangular cross-section with constant width b and variable depth d, the relationship $I(x) = cA(x)^3$ is satisfied since $I(x) = b\, d(x)^3 / 12 = A(x)^3 / 12b^2$. The dimensions of the column are the same as in Case 1 but in this case the depth d is allowed to change in the range of 5 mm to 125 mm in steps of 5 mm instead of the width b. Again 25 different cross-sectional areas are allowed in the design.

When the column is simply supported at both ends, the buckling optimization of the column represents a single modal problem. Using the evolutionary procedure described in the last section, the optimum shapes of the column for the above three cases are obtained after between 12 and 20 iterations. These shapes are given in Figure 6.1. Due to symmetry only a quarter of each column is shown.

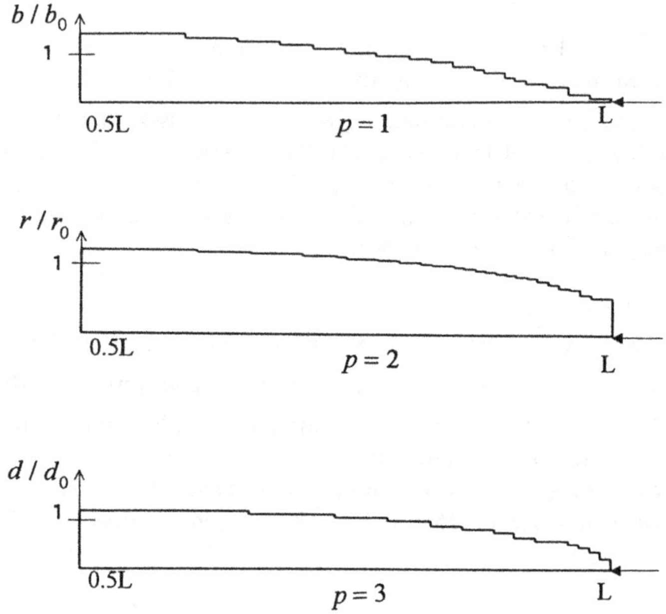

Fig. 6.1. Optimal shapes of simply supported columns.

The evolutionary histories of the buckling load factors for the column are shown in Figure 6.2. In Table 6.1 the final optimum buckling load factors obtained by the ESO method are compared with the exact solutions (see Haftka and Gürdal 1992). Excellent agreements between the two are observed.

Table 6.1. Optimum buckling load factors for simply supported columns

p	$\lambda_1 / \lambda_1^{uniform}$		%
	Exact	ESO	difference
1	1.215	1.214	- 0.1
2	1.333	1.328	- 0.4
3	1.407	1.400	- 0.5

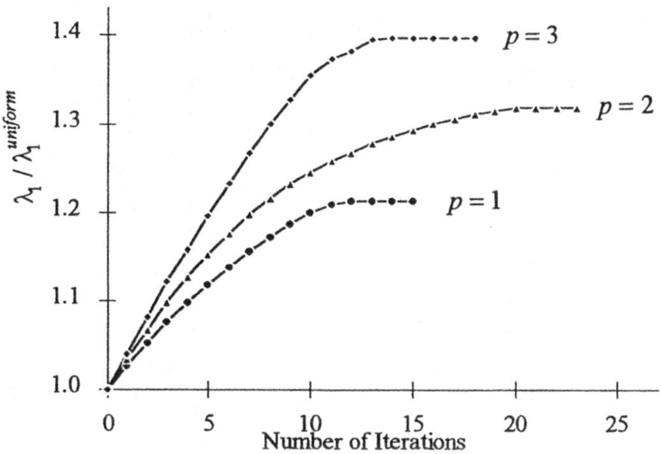

Fig. 6.2. Optimization histories for the simply supported columns.

6.6 Example of Frame Optimization

A three member portal frame which was analyzed by Szyszkowski *et al.* (1989) for bimodal buckling is considered. The frame is pinned at the base and is statically indeterminate. The frame layout and the loading are shown in Figure 6.3. All the three members are of circular cross-sections and of equal length of 1 m. r^2 is initially equal to 30 mm^2 and is allowed to vary to the maximum of 50 mm^2 and to the minimum of 10 mm^2 in steps of 1 mm^2. Each member is divided into 10 elements of equal length. For the frame with a uniform cross-section, the first buckling mode is anti-symmetric with sway and the second buckling mode is symmetric without sway. High compressive loads are applied in the horizontal direction to make the problem bimodal from the outset. Applying the bimodal method by using equation (6.11) for the calculation of sensitivity numbers, we obtain the optimum shape of the frame as shown in Figure 6.3. The corresponding buckling load is 1.247 times that of the uniform frame. If, however, we only consider the first buckling mode and use equation (6.6) instead, the buckling load factor can only reach 1.125 times that of the uniform frame. The evolutionary histories of the first two eigenvalues using both single and bimodal methods are given in Figure 6.4.

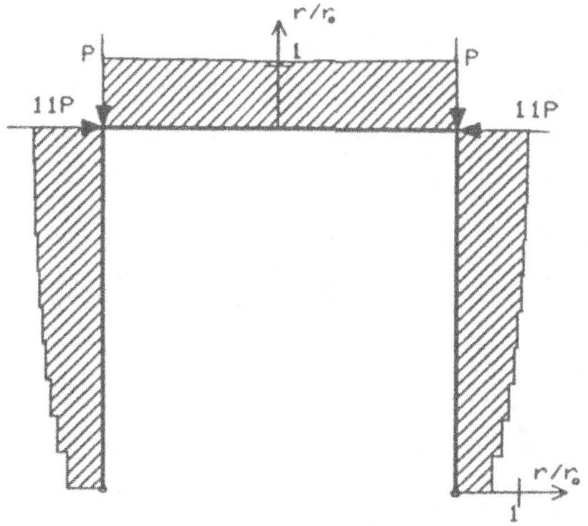

Fig. 6.3. Optimal shapes of the portal frame.

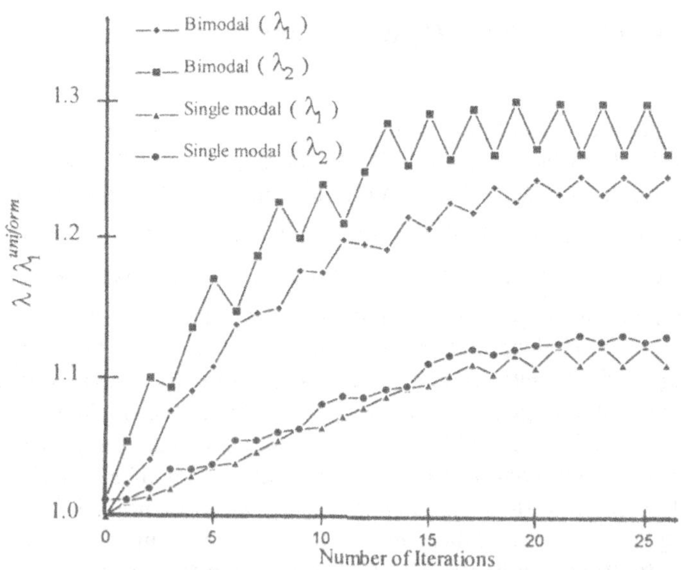

Fig. 6.4. Comparisons of eigenvalues from single modal and bimodal methods.

6.7 Plate Optimization

In this section we discuss the buckling optimization of a square plate, which is simply supported on the four sides and is subjected to *constant uniaxial force* as shown in Figure 6.5. This loading condition is to be differentiated from that of *constant uniaxial stress*. As the plate thickness is variable, these two loading conditions are significantly different and so are the corresponding optimum solutions (Pandey and Sherbourne 1992).

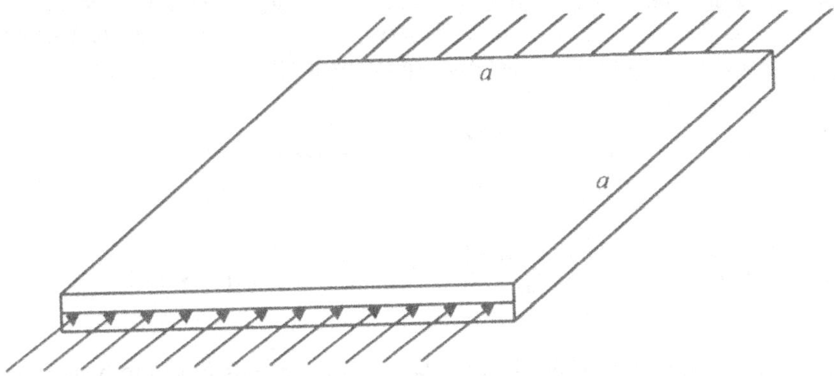

Fig. 6.5. Uniaxially loaded square plate.

center

9	14	18	18	19	20	21	22	23	23
6	12	15	16	18	20	21	22	22	22
5	6	12	15	16	18	20	21	21	22
13	5	9	13	15	16	18	19	20	20
16	15	5	9	13	15	16	18	19	19
19	16	5	5	9	13	15	15	17	17
15	15	8	5	5	9	12	14	15	16
25	25	15	15	15	5	8	11	13	15
25	25	25	18	15	15	5	5	10	12
25	25	25	15	15	15	5	5	5	8

corner

Fig. 6.6. Optimal thickness distribution for a quarter of the plate.

We apply the evolutionary procedure to the square plate of dimensions 1m x 1m. The plate is divided into 20 × 20 quadrilateral elements. Eight node isoparametric elements are used in the finite element modelling. The initial uniform thickness of the plate is equal to 15 mm and the thickness t is allowed to vary to the maximum of 25 mm and to the minimum of 5 mm in steps of 1 mm. The final thickness distribution obtained from the evolutionary procedure is given in Figure 6.6 for a quarter of the plate. The buckling load for this plate is 1.47 times that of the uniform plate of the same weight.

It is observed from the result shown in Figure 6.6 that material is redistributed to the centre and the four corners of the plate and a hinge line is formed between the centre and the corners. From these observations a smooth shape for the optimum plate is proposed using the following equations for the thickness:

$$t(r) = kt_{\min} - (k-1)t_{\min}\left(\frac{r}{a/2}\right)^2 \qquad \text{when } r \le a/2, \tag{6.13a}$$

$$t(r) = kt_{\min} - (k-1)t_{\min}\left(\frac{\sqrt{2} - 2r/a}{\sqrt{2}-1}\right)^2 \qquad \text{when } r > a/2, \tag{6.13b}$$

where a is the length of the square plate, r is the radius from the plate centre and $k = t_{\max} / t_{\min}$.

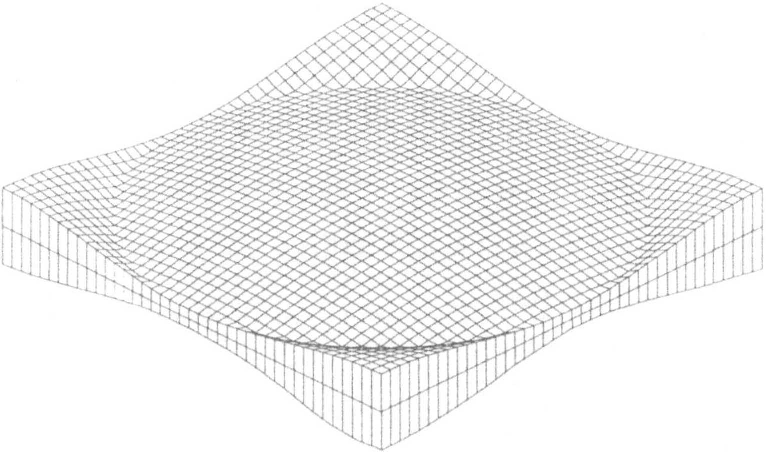

Fig. 6.7. Proposed optimal shape for the simply supported square plate ($k = 4$).

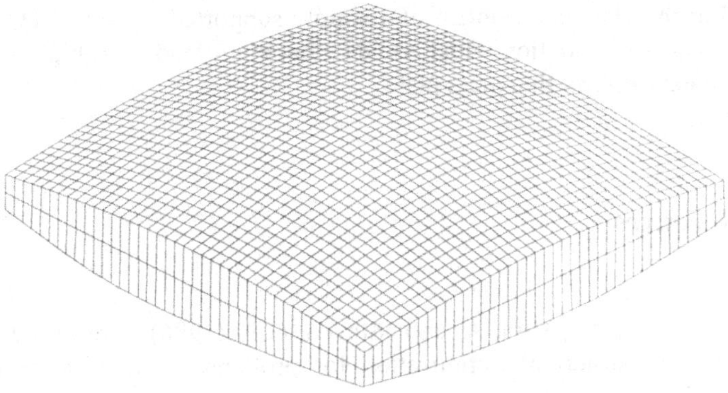

Fig. 6.8. Optimal shape for the plate by Pandey and Sherbourne (1992).

The shape of the plate as defined by the above equation is shown in Figure 6.7. Finite element analysis is carried out for this plate using 40 × 40 eight node isoparametric elements. The buckling load factors obtained for different values of *k* are tabulated in Table 6.2. It is seen that the highest buckling load occurs when *k* is equal to 4 and it is 1.37 times that of the uniform thickness plate. This is a significant improvement on the shape presented by Pandey and Sherbourne (1992). The buckling load of the Pandey and Sherbourne plate, shown in Figure 6.8, is 1.28 times that of the uniform thickness plate.

Table 6.2. Buckling load factors for the plate as defined by equation (6.13)

k	3.0	3.5	4.0	4.5	5.0
$\lambda_1 / \lambda_1^{uniform}$	1.355	1.369	1.370	1.358	1.331

6.8 Conclusion

In this chapter we have extended the ESO method to buckling optimization. By ignoring the change in the stress matrix, the sensitivity number can be easily calculated. However, such a sensitivity number is not applicable to optimization involving element removal. For this reason we have restricted the discussion in the chapter to cross-sectional variations.

The evolutionary procedure works well for the optimization of columns and frames. For the classical problem of a simply supported square plate, we obtain a novel thickness distribution which yields a buckling load 37% higher than that of the equivalent uniform thickness plate.

6.9 References

Chu, D.N., Xie, Y.M., Hira, A. and Steven, G.P. (1996) On various aspects of evolutionary structural optimization for problems with stiffness constraints. *Finite Elements in Analysis and Design* (to appear).

Haftka, R.T. and Gürdal, Z. (1992) *Elements of Structural Optimization*, 3rd edn., Kluwer Academic Publishers, Dordrecht.

Haug, E.J. (1982) Optimization of structures in which repeated eigenvalues occur. In: A.J. Morris (ed.), *Foundations of Standard Optimization: A Unified Approach*, John Wiley & Sons, New York, 559-572.

Keller, J.B. (1960) The shape of the strongest column. *Arch. Ration. Mech. Anal.* **5**(4), 275-285.

Manickarajah, D., Xie, Y.M. and Steven, G.P. (1995) A simple method for the optimization of columns, frames and plates against buckling. In: S. Kitipornchai, G.J. Hancock and M.A. Bradford (eds.), *Structural Stability and Design*, Balkema, Rotterdam, 175-180.

Manickarajah, D., Xie, Y.M. and Steven, G.P. (1996) On optimum design of plate structures against buckling. *Struct. Optim.* (submitted).

Olhoff, N. and Rasmussen, S.H. (1977) On single and bimodal optimum buckling loads of clamped columns. *Int. J. Solids Struct.* **13**, 605-614.

Pandey, M.D. and Sherbourne, A.N. (1992) Mechanics of shape optimization in plate buckling. *J. Eng. Mech. ASCE* **118**(6), 1249-1266.

Rodrigues, H.C., Guedes, J.M. and Bendsøe, M.P. (1995) Necessary conditions for optimal design of structures with a nonsmooth eigenvalue based criterion. *Struct. Optim.* **9**, 52-56.

Simitses, G.J., Kamat, M.P. and Smith, C.V., Jr. (1973) Strongest column by the finite element displacement method. *AIAA J.* **11**(9), 1231-1232.

Spillers, W.R. and Levy, R. (1990) Optimum design for plate buckling. *J. Struct. Eng. ASCE* **116**(3), 850-858.

Szyszkowski, W. and Watson, L.G. (1988) Optimization of the buckling load of columns and frames. *Eng. Struct.* **10**, 249-256.

Szyszkowski, W., Watson, L.G. and Fietkiewicz, B. (1989) Bimodal optimization of frames for maximum stability. *Comput. Struct.* **32**(5), 1093-1104.

Tadjbakhsh, I. and Keller, J.B. (1962) Strongest columns and isoperimetric inequalities for eigenvalues. *J. Appl. Mech.* **29**, 159-164.

Chapter 7 ESO for Pin- and Rigid-Jointed Frames

Evolutionary Structural Optimization (ESO) has been previously successfully applied to 2D and 3D optimization of continuous domains for static, dynamic and stability analyses. In the preceding chapters attention has been focussed on shape and topology optimization under different optimality constraints. There is a significant class of practical structural forms involving frame structures where the layout of the structure is known and only the size of the frame members are design variables. The frames can be either pin-jointed or rigid-jointed or a combination of the two. Finding the appropriate combination of member sizes that meet some optimality objectives under single or multiple load environments is the subject of this chapter. Also for frame structures this chapter shows that topology optimization can be conducted and is successful, provided the initial design space or ground structure is sufficiently large.

7.1 Introduction

This chapter considers the application of the ESO concept to the size and topology optimization of frames both pin-jointed and rigid-jointed. Size optimization is the earliest form of structural optimization whereby aspects of the cross sectional size of structural members are adjusted to give better structural performance. The size variation may be continuous or discrete. If the sizes of elements are allowed to go to zero then they are removed and the final optimum is thereby a combination of topology and size optimization. This introductory presentation is focused on stress optimality constraints, but the algorithms can be easily adapted to stiffness, frequency and other constraints.

If the structure is statically determinate to begin with and the objective is a fully stressed design, the minimum weight is simply determined by the given design target stress. If the structure has redundant members but statically determinate support conditions then the optimum sizes and topology are such that certain member areas go to zero to produce a statically determinate structure (Rozvany 1989). For size optimization the layout of the structure is sometimes initially determined and therefore the final design is not necessarily the best topological layout.

For other problems if there is a lot of redundancy in the structure, topology optimization will successively remove lowly stressed members and re-analyze the structure in the same manner as the continuum ESO of Chapter 2.

The seminal work of Michell (1904) shows that for an optimum fully stressed design, the members must form a set of curved lines such that the tension members always intersect the compression members at right angles, ie. they form a set of curvilinear coordinates. Many examples of Michell fully stressed pin-jointed frames are given in the book by Hemp (1973).

For static analysis there are two distinct branches of optimization, either the target is a fully stressed design (FSD) where all members are at their maximum allowable stresses, albeit different in tension and compression, and varying from one member to another; or it is a minimum mass for a given stiffness. The latter generally comes in the form of prescribed displacement or displacements. For pin-jointed or rigid-jointed frame analysis an additional constraint in either of these two forms can be that, in compression, there is no Euler buckling .

This chapter presents the overall iterative approach of the ESO algorithm for pin- and rigid-jointed frame structures. The process is initially applied to the classic ten bar truss as discussed in Fleury (1979) and Haftka and Gürdal (1992). For this traditional problem the stress limits on all bars are the same except for one member where the stress limit is gradually increased to three times the value of the other members. Other examples of 2D size optimization are presented for Michell type structures where the initial structural form can be very far removed from the optimum but for which the ESO algorithm simply seeks the correct configuration. Although the optimization of 3D pin-jointed frames would appear to be more demanding (Pederson 1993), the ESO algorithm efficiently finds the best configuration regardless of the complexity and redundancy of the initial topology.

Finally the ESO algorithm is applied to rigid-jointed frames with similar success.

A feature of the work presented in this chapter is the simplicity of the ESO process for either topology and/or size optimization of frame structures. Generally the computer code is a few hundred lines and adapts easily to whatever problem is being solved, 2D or 3D, pin- or rigid-jointed frames and combinations. A sample FORTRAN code for 2D frame size optimization is presented as an appendix at the end of this chapter.

7.2 ESO Algorithm for Pin-Jointed Frames

1. An initial structure is defined including loads and support conditions.
2. Conduct FEA.
3. Calculate member stresses.
4. Compare member stress with target value.
5. If absolute stress is above target, increase area by a small increment; or if absolute stress is below target, decrease area by a small increment.
6. If area is diminished to zero, remove element from structure; or if area reached prescribed lower or upper bound then freeze area.
7. Check to see if previously frozen areas need unfrozen.
8. If volume change per iteration is within a small convergence tolerance or a prescribed iteration limit has been reached, then stop and print results; if not, go to 2.

The area step has to be very small of the order (original area)/1000 or less. The member target stresses can be different for each member and different for tension and compression.

7.3 ESO Algorithm for Rigid-Jointed Frames

1. An initial structure is defined including loads and support conditions.
2. Conduct FEA.
3. Calculate member maximum combined bending and axial stresses.
4. Compare member stress with target value.
5. If absolute stress is above target, increase geometric property by a small increment; or if absolute stress is below target, decrease geometric property by a small increment.
6. If geometric property diminished to zero remove element from structure;

or if property reached some prescribed lower or upper bound then freeze it.

7. Check to see if previously frozen member sizes need unfrozen.

8. If volume change per iteration is within a small convergence tolerance or a prescribed iteration limit has been reached, then stop and print results; if not, go to 2.

The geometric property changed by ESO could be one of several from width, depth, radius for circular section, etc.

7.4 Examples of Size Optimization of Pin-Jointed Frames

7.4.1 The Ten Bar Truss

Figure 7.1 shows the truss of ten bars fixed in both horizontal and vertical directions at the wall on the left hand side. Two point loads of equal value are applied at nodes 2 and 4. The material has a nominal Young's modulus of 1.0 and initial area of 5.

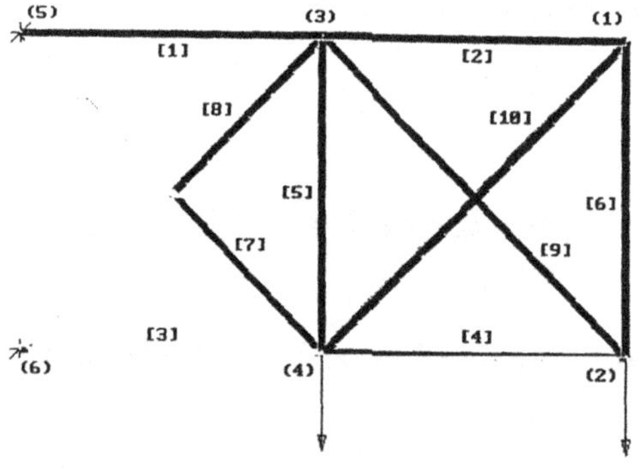

Fig. 7.1. Ten bar truss for size optimization.

Two sets of test analyses are carried out, one with a minimum allowable area of 0.1 and the other with a minimum of 0.0. The practice of having a minimum area of 0.1 is unclear as it seems better to let member disappear rather than artificially maintain a small area. It can only be surmised that this practice initiated with software that could not cope with members being removed. An alternative reason for not allowing a member to be removed is the possibility of it being needed later in the optimization iterations. In the ESO process this possibility is catered for by setting the lower size bound at, say 10^{-6} of the original and allowing for unfreezing as indicated in step 7 of the above ESO algorithms.

This example proves interesting for two reasons. Firstly when the stress limit is increased in one of the members, namely member 9 in Figure 7.1, logic would say that the ability of that bar to sustain higher stress would lead to a lighter structure regardless of the lower limit in member area. This turns out to be not the case for both sets of test data. With or without a minimum area there comes a point at which the achievable optimum structure has an increase in volume and a fully stressed design is not achieved and the stress in bar 9 cannot keep up with the allowable. This unusual behaviour has been reported by many researchers solving this problem. Here, by using ESO, some explanation of the reason for this can be offered. Tables 7.1 and 7.2 track the member stresses and areas for a range of allowable stresses in bar 9. With a zero minimum area in Table 7.1 it is seen that as the allowable stress in bar 9 increases the form of the optimum frame totally changes and the area of bar 9 goes to zero, thus not being able to sustain any stress (see last column in the table). Thus it can be said that this test is not logical as bar 9 is not part of the structure.

Even when there is a minimum area, as in Table 7.2, the size of bar 9 goes to this minimum and the rest of the structure does not direct sufficient load to this bar to make its stress go any higher. Another way of looking at this is to plot the final volume against the allowable stress in bar 9 as shown in Figure 7.2. From this it is clear that the expected reduction is happening up to a stress of 37000. Thereafter the volume goes up as the area of bar 9 goes down towards zero and the load finds other paths.

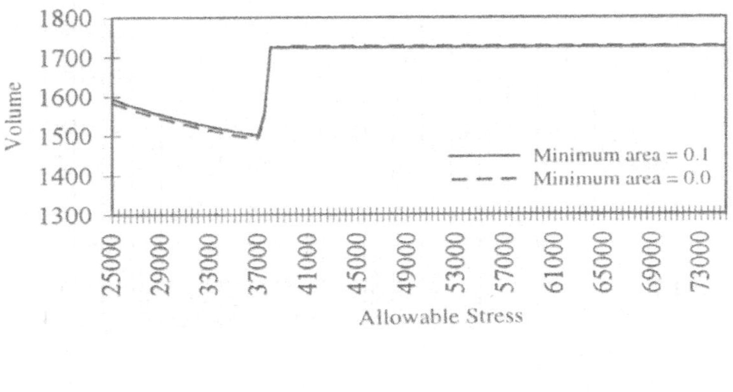

Fig 7.2. Final volume of the ten bar truss against allowable stress in bar 9.

Table 7.1. Ten bar truss, member areas and stresses for various stress limits in bar 9, minimum allowable area = 0.0.

Stress Limit	25000	37100	37300	37500	37600	37700	37900	40000	50000	75000
Area 1	8.000	7.955	7.911	6.925	4.285	4.144	4.073	4.009	3.999	4.000
Stress 1	25000	25002	25002	24998	24994	24994	24993	24994	24999	25000
Area 2	0.000	0.045	0.089	1.075	3.715	3.855	3.927	3.991	3.999	4.000
Stress 2	0	24591	24768	25008	25006	25006	25006	25005	24999	25000
Area 3	8.000	8.045	8.089	9.075	11.715	11.855	11.926	11.991	11.999	12.000
Stress 3	25000	24997	24997	25000	25001	25002	25002	25001	24999	25000
Area 4	4.000	3.955	3.911	2.925	0.285	0.145	0.073	0.009	0.000	0.000
Stress 4	25000	25004	25005	24996	24917	24833	24632	22897	0	0
Area 5	0.000	0.000	0.000	0.000	0.000	0.000	0.000	0.000	0.000	0.000
Stress 5	0	0	0	0	0	0	0	0	0	0
Area 6	0.000	0.045	0.089	1.075	3.715	3.855	3.927	3.991	3.999	4.000
Stress 6	0	24609	24774	25008	25006	25006	25006	25005	24999	25000
Area 7	5.657	5.721	5.783	7.177	10.911	11.109	11.211	11.301	11.312	11.313
Stress 7	25003	25003	25004	25000	25001	25001	25001	25000	24999	25000
Area 8	5.657	5.593	5.531	4.137	0.403	0.205	0.103	0.013	0.000	0.000
Stress 8	25003	24994	25003	24996	24945	24890	24784	23040	0	0
Area 9	5.657	3.769	3.707	2.759	0.269	0.136	0.067	0.009	0.000	0.000
Stress 9	25003	37107	37289	37504	37605	37702	37910	40520	0	0
Area 10	0	0.065	0.127	1.519	5.235	5.451	5.553	5.643	5.657	5.657
Stress 10	0	25002	25021	24989	24998	24998	24998	24999	25003	25002
Mass	1583.8	1492.7	1494.3	1552.4	1710.8	1719.2	1723.5	1727.4	1727.9	1727.9

Table 7.2. Ten bar truss, member areas and stresses for various stress limits in bar 9, minimum allowable area = 0.1.

Stress Limit	25000	35000	36900	37100	37200	37250	37300	37500	37700	37900	38100	40000	75000
Area 1	7.937	7.909	7.903	7.901	7.901	7.901	7.81	6.918	4.282	4.106	4.106	4.106	4.106
Stress 1	25000	25000	25000	25000	24997	25002	25000	25001	25000	25000	24995	24995	24995
Area 2	0.100	0.100	0.100	0.100	0.100	0.100	0.19	1.082	3.718	3.894	3.894	3.894	3.894
Stress 2	15553	22559	24405	24667	24715	24756	25000	24983	25000	25002	25001	25001	25001
Area 3	8.063	8.091	8.097	8.099	8.099	8.099	8.19	9.082	11.718	11.894	11.894	11.894	11.894
Stress 3	25000	25001	25002	25001	25002	24997	25000	25001	25003	25000	25000	25000	25000
Area 4	3.739	3.909	3.903	3.901	3.901	3.901	3.81	2.91	0.282	0.106	0.106	0.106	0.106
Stress 4	25000	24998	25002	24995	24994	25006	25000	25012	25000	24694	24695	24695	24695
Area 5	0.100	0.100	0.100	0.100	0.100	0.100	0.1	0.1	0.100	0.100	0.100	0.100	0.100
Stress 5	1.140	11.000	16.000	12.47	18	19	13.3	23.1	12.2	100.3	129	129	129
Area 6	0.100	0.100	0.100	0.100	0.100	0.100	0.19	1.082	3.718	3.894	3.894	3.894	3.894
Stress 6	15537	22564	24409	24672	24719	24761	25002	24983	25004	25000	25000	25000	25000
Area 7	5.745	5.785	5.795	5.797	5.797	5.797	5.922	7.19	10.918	11.162	11.162	11.162	11.162
Stress 7	25003	25001	24996	25000	25002	24995	25003	25005	25000	25000	24996	24996	24996
Area 8	5.569	5.529	5.519	5.517	5.517	5.517	5.39	4.126	0.398	0.15	0.152	0.152	0.152
Stress 8	25004	24996	25003	25000	24995	25002	25004	24984	25000	24933	24933	24933	24933
Area 9	5.569	3.949	3.739	3.716	3.707	3.703	3.614	2.75	0.262	0.100	0.100	0.100	0.100
Stress 9	25004	34997	36909	37100	37196	37255	37311	37485	37504	37716	37716	37716	37716
Area 10	0.100	0.127	0.139	0.139	0.139	0.141	0.266	1.532	5.262	5.506	5.506	5.506	5.506
Stress 10	21969	24926	25001	24919	24967	25009	24801	24996	25000	24995	24995	24995	24995
Mass	1593.0	1511.1	1500.6	1499.7	1499.1	1498.8	1504	1556.8	1714.6	1725.3	1725.3	1725.3	1725.3

From Figure 7.2 it is seen that there is almost a step change in volume at stress limit in bar 9 of about 37000. Thus it can again be concluded that there are two different structures when bar 9 is there or is not there. It should be noted that if the starting structure does not include member 9 then the ESO process ends with the shape of Figure 7.3(c).

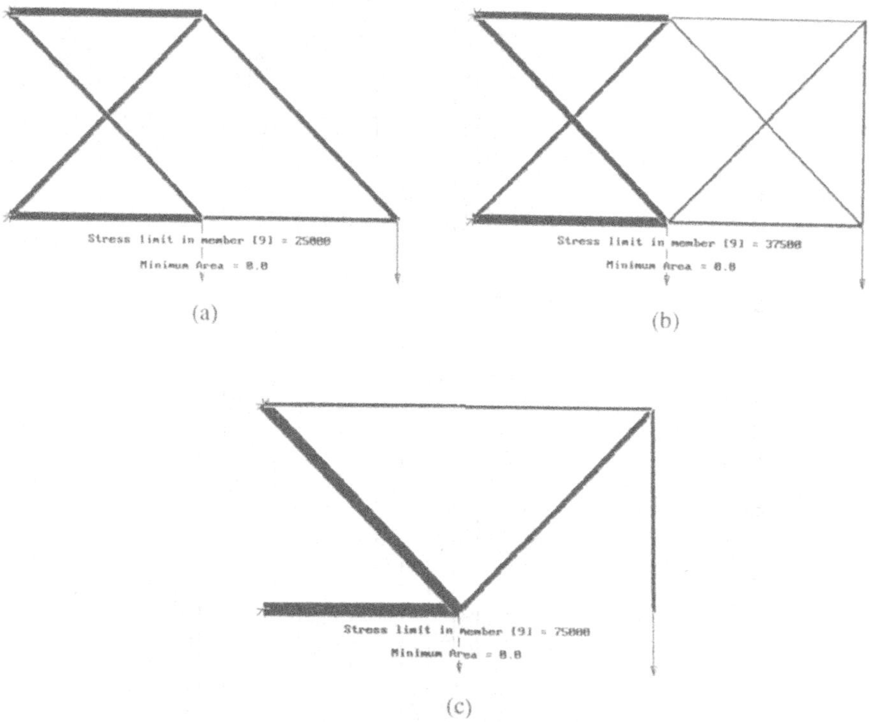

Fig 7.3. Final optimum structures for three different stress limits in bar 9.

7.4.2 Michell Arch

This second pin-jointed frame example is based on the known solution for the optimum frame to support a central load between roller supports. The Michell solution for this is a semi-circular arch with a finite number of spokes which for a fully stressed design has a volume of

$$V = \frac{2nLW}{\sigma}\tan\left(\frac{\pi}{2n}\right), \tag{7.1}$$

where W is the applied load, σ is the allowable (or target) stress, n is the number of segments of the arch and L is the length of half span which in this case is equal to the radius of the arch R.

Here a six segment arch is studied. The length $L = 10$, the load $P = 1$ and the Young's modulus is set at 1 as is the target stress σ. Figure 7.4 shows the six segment arch and Table 7.3 shows the results of member stresses after the ESO process for a range of area increments with a starting area of 0.5.

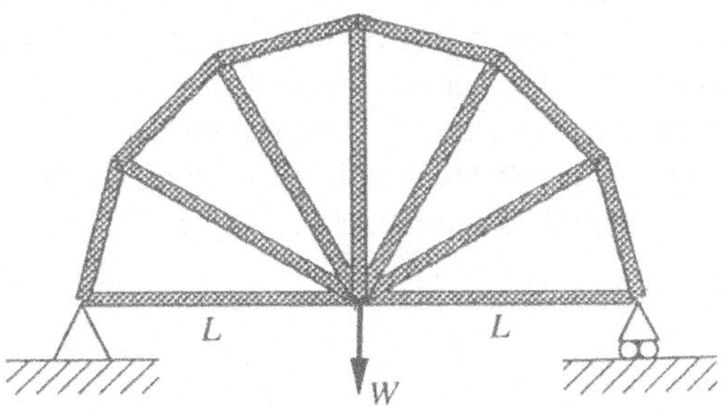

Fig. 7.4. A six segment Michell arch on roller supports.

Table 7.3. ESO results for six segment arch with various area increments.

Area increment	Rim member stress	Inside spoke stress	Outside spoke stress	Final arch volume	Error in final volume	RMS DEV
0.01	0.995458	1.030572	0.956964	31.95030	0.63321%	0.74100
0.001	0.999301	0.999809	0.999814	32.16818	0.04440%	0.00993
0.0001	1.000074	0.999809	0.999813	32.15576	0.00578%	0.00448
0.00001	0.999996	1.000033	0.999963	32.15380	0.00032%	0.00075

In this table the column titled RMS DEV gives the sum of the square of the normalized deviation from the target stress at the iteration limit and can be seen to behave in much the same way as the error in the volume.

Several observations can be made from Table 7.3. Firstly the evolutionary process always converges to the correct result. The error in the area can be either side of the exact value as can the stresses. This occurs because of the nature of the switch in the algorithm, whereby if the stress is over target then the area is increased and if the stress is below target then the area is reduced. At the time the iteration limit is reached it could be that some areas are below their optimum value and some are above. Alternatively all areas could be above or they could all be below, the final volume may converge from above or below.

This example starts with a statically determinate structure and clearly ends with the same number of members. As a simple extension, two additional starting structures are investigated, these are shown below in Figure 7.5. One has 27 bars fitting every possible connection for the eight nodes and the other more involved starting position has 35 bars and two extra nodes. For both of these starting structures two runs are conducted with different iteration limits and area increments. Regardless of the starting structure the final optimum structures have only the six segment Michell arch with all extraneous members removed. The results are shown in Table 7.4 below.

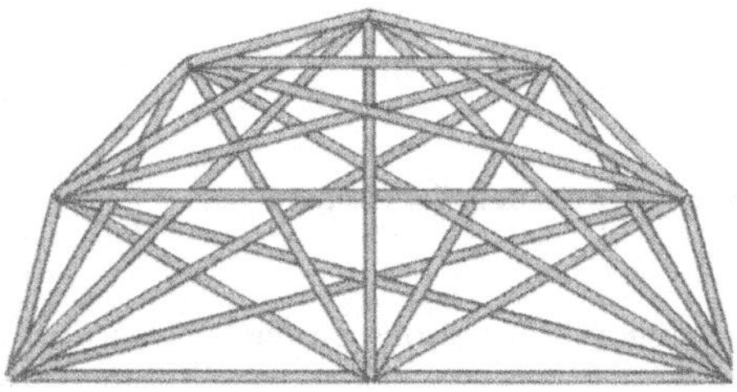

(a)

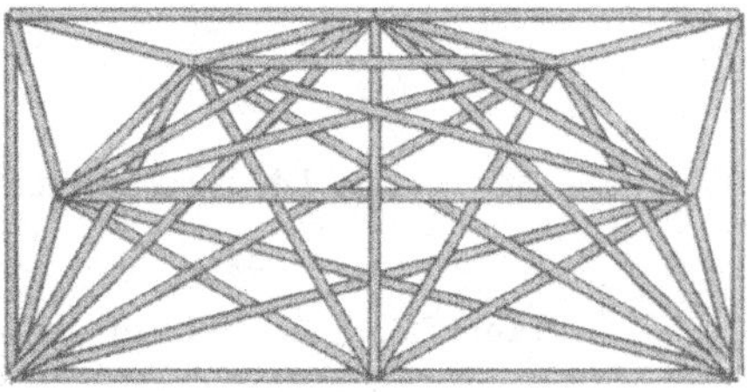

(b)

Fig. 7.5. Two highly redundant starting structures with fixed supports.

Table 7.4. Results for starting structures shown in Figure 7.5.

Starting number of bars	Area increment	Final volume	Error in volume	RMS DEV
27	0.001	32.16818	0.04440%	0.00993
27	0.00001	32.15380	0.00032%	0.00075
35	0.001	32.16818	0.04441%	0.00993
35	0.00001	32.15380	0.00032%	0.00075

It is interesting to note that although the starting structures are highly statically indeterminate the ESO process always produces statically determinate optima if the support conditions are statically determinate.

Next we consider a similar arch structure but with the two supports changed to fixed conditions. Two starting structures with totally fixed supports are shown in Figure 7.6.

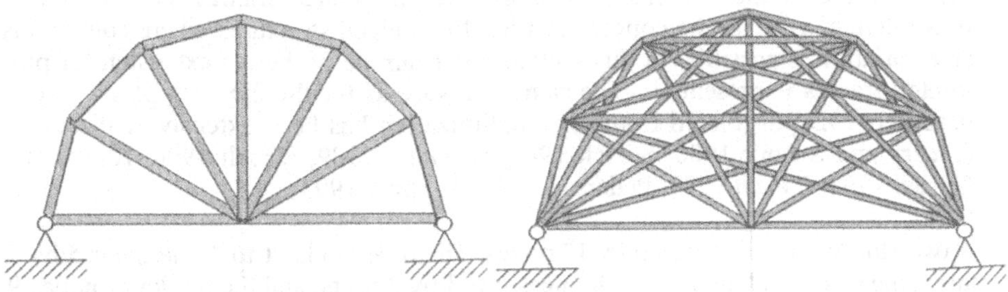

(a) (b)

Fig. 7.6. Two starting structures with totally fixed supports.

Rather than tabulating all the results for individual optimization runs only the final results for area increment of 0.00001 with starting area of 0.5 will be given. Figure 7.7(a) is the final topology for the initial structure of Figure 7.6(a) with a final volume of 29.47420 and similarly Figure 7.7(b) is the final topology for the initial structure of Figure 7.6(b) with a final volume of 28.03861.

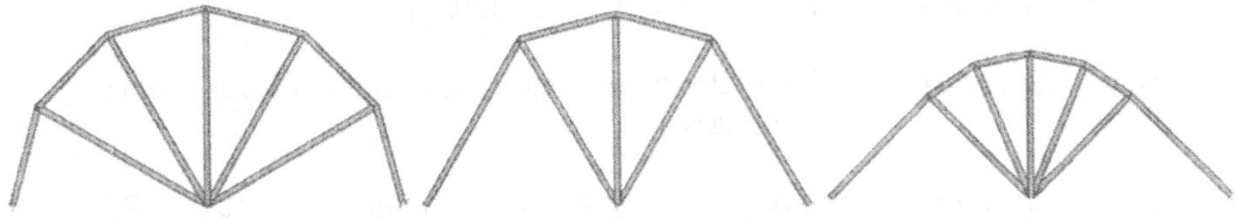

(a) (b) (c)

Fig. 7.7. Final structures for an arch with fixed supports.

In Figure 7.7 it is interesting to note that despite the initial structural form, the evolutionary structural optimization has taken the optimum in the direction of the corresponding Michell structure for an arch with fixed supports. This is discussed in Hemp (1973) and is shown in Figure 7.7(c), of which the volume is $(1+\pi/2)WL/\sigma = 25.7079$. From Figures 7.7(a) and 7.7(b) it is seen that given any form of the starting structure the ESO algorithm heads towards a final form that most closely resembles the exact optimum. Had the starting structural topology included this Michell optimum the ESO process would have found this.

7.4.3 Three Dimensional Pin-Jointed Frames

The above examples in this chapter on 2D pin jointed frames have served to illustrate and validate the concept of ESO for skeletal structures where continuous size variation is possible. In this section one example of the 3D extension for pin-jointed frames is presented. The same software as for the 2D examples above is used here. 3D pin jointed frame size optimization has been extensively discussed (Fleury and Schmit 1980; Kirsch 1989; Rozvany 1989; Kirsch 1990; Haftka and Gürdal 1992; Kirsch 1993; Pederson 1993; Topping 1993).

A two storey tower is defined by 12 nodes. The base, nodes 1 to 4, measures 5 by 5 units; the next level, nodes 5 to 8, measures 3 by 3 units; and the top level, nodes 9 to 12, is 2 by 2 units. The storeys are 5 units vertically separated. The base nodes are fixed in all three directions and the following loads are applied: $F_{5,x} = F_{5,y} = F_{5,z} = F_{12,x} = 1.0$, $F_{8,z} = F_{10,y} = 2$, $F_{6,y} = -1$.

Four possible levels of starting structure could be constructed from the 12 nodes, these are designated as Level 1 Structure to Level 4 Structure and they are composed as follows:

Level 1 Structure — All possible horizontal and vertical members and all external single level X bracing = 36 bars.

Level 2 Structure — Level 1 + all floor X bracing = 42 bars.

Level 3 Structure — Level 2 + all internal single and two level X bracing = 54 bars.

Level 4 Structure — Every possible interconnection = 11 + 10 + ... 2 + 1 = 66 bars.

The evolutionary process starts with each bar having an area of 1 and a Young's modulus of 1. The area increment allowed is 0.001, and the ESO process converged in a straight forward manner. For each of the above four structural levels the start and finish topologies are shown in Figure 7.8. Various convergence strategies have been tested such as area increment halving from a large initial starting jump; area increments in proportion to the ratio of actual stress over target stress and others. All logical choices have resulted in convergence. The code for this example can be obtained by modifying the FORTRAN program in the Appendix to this chapter.

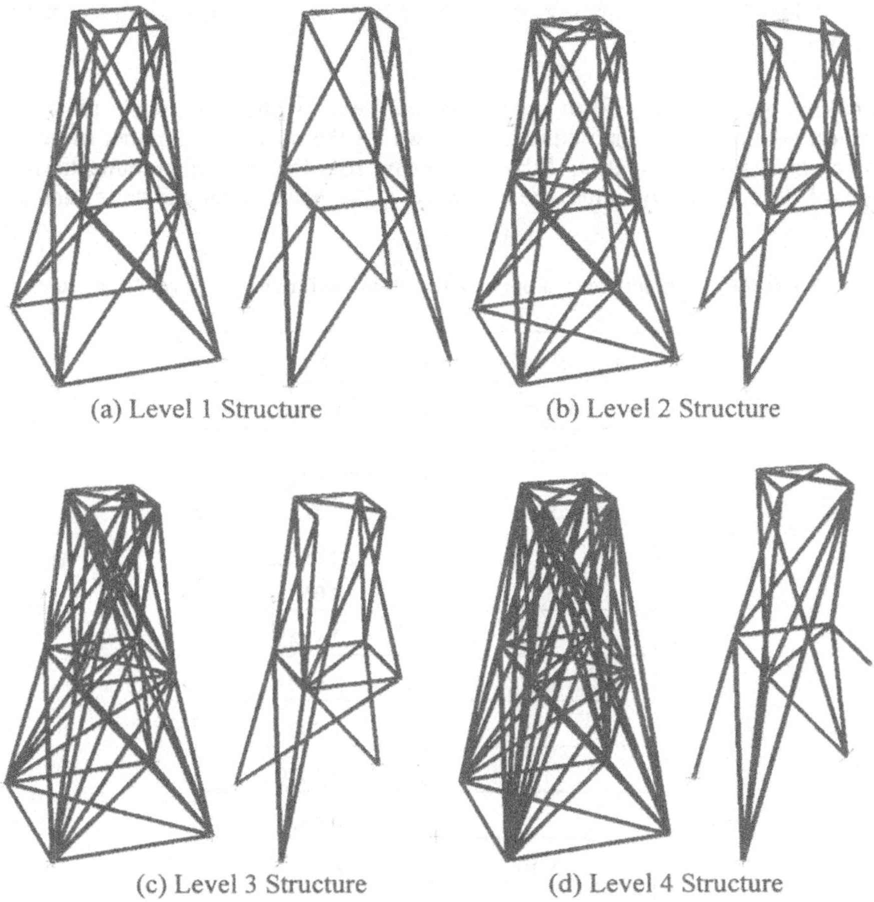

Fig. 7.8. Start and finish topologies for size optimization of four levels of 3D frames.

Table 7.5. Volumes of optimal designs for the four starting levels of 3D frames.

Level	Volume
1	152.9415
2	100.9360
3	98.9198
4	95.6711

The volumes of the optimal designs for the four starting levels are given in Table 7.5 above.

Looking back it is logical to expect that the Level 4 Structure would provide the lowest volume optimal structure as it allows for the maximum number of possible topologies. For information Table 7.6 below lists the final interconnections and areas for the Level 4 optimum. Each member is denoted by their end node numbers.

Table 7.6. Member areas of optimal 3D pin-jointed frame for Level 4 Structure.

Member end nodes	Area
6-7	1.0000
7-8	0.3107
8-1	0.4687
10-11	0.0515
11-12	0.0515
9-12	0.6185
1-5	2.0545
3-7	1.7907
4-8	0.2079
1-8	0.4687
5-10	0.3367
7-10	1.9341
8-9	1.1571
5-12	0.3367
5-7	0.6831
9-11	0.0729
10-12	1.1505
3-5	0.5691
8-10	0.0565
1-9	1.0537
1-10	1.5721
3-12	0.3211

7.5 Topology Optimization of Pin-Jointed Frames

Topology optimization has a different algorithm from size optimization. The design space is populated with a regular mesh of points to a prescribed density. The nodal mesh is then connected with patterns of pin-jointed elements. The pattern can be anything from every possible connection, i.e. for n nodes there is $(n - 1) + (n - 2) + \dots 1$ bars, to patterns of stars. After FEA, bar elements with stress levels less than the Rejection Ratio times the maximum stress are eliminated. This cycle is repeated along the identical lines to that of the ESO procedure for continuum analysis.

The paper by Sankaranarayanan *et al.* (1993) presents several solutions to this topic using the Simultaneous Analysis and Design (SAND) approach whereby the design variables and the analysis variables are part of the same single level optimization iteration rather than doing analysis the optimization separately. The modern version of SAND is described in Haftka and Gürdal (1992). The Optimality Criteria (OC) method developed and extensively researched by Rozvany, has also been applied many times to problems of topology optimization for pin-jointed structures (Rozvany 1989; Rozvany *et al.* 1994; Rozvany *et al.* 1995). The paper by Topping (1993) contains a large bibliography on this topic.

To illustrate this technique two examples are presented. First a stress based ESO process for 2D pin-jointed frame is demonstrated. The second example is a 3D structure optimized under a displacement constraint.

Figure 7.9(a) shows a typical coarse form of starting grid with a 2 by 1 design space divided into a 10 by 5 mesh with each cell cross-braced. The supports at bottom left and right are fixed and a central load is applied. Figure 7.9 traces the evolution of the design space as lowly stressed members are removed and the remainder moves towards being fully stressed. Given the starting topology it is noteworthy that the final evolved frame has a resemblance to the Michell one shown in Figure 7.7(c).

The above approach to topology optimization will only work if there are a very large number of bars in the initial structure. The tendency to have a regular grid limits the options for finding the optimum topology within the initial grid.

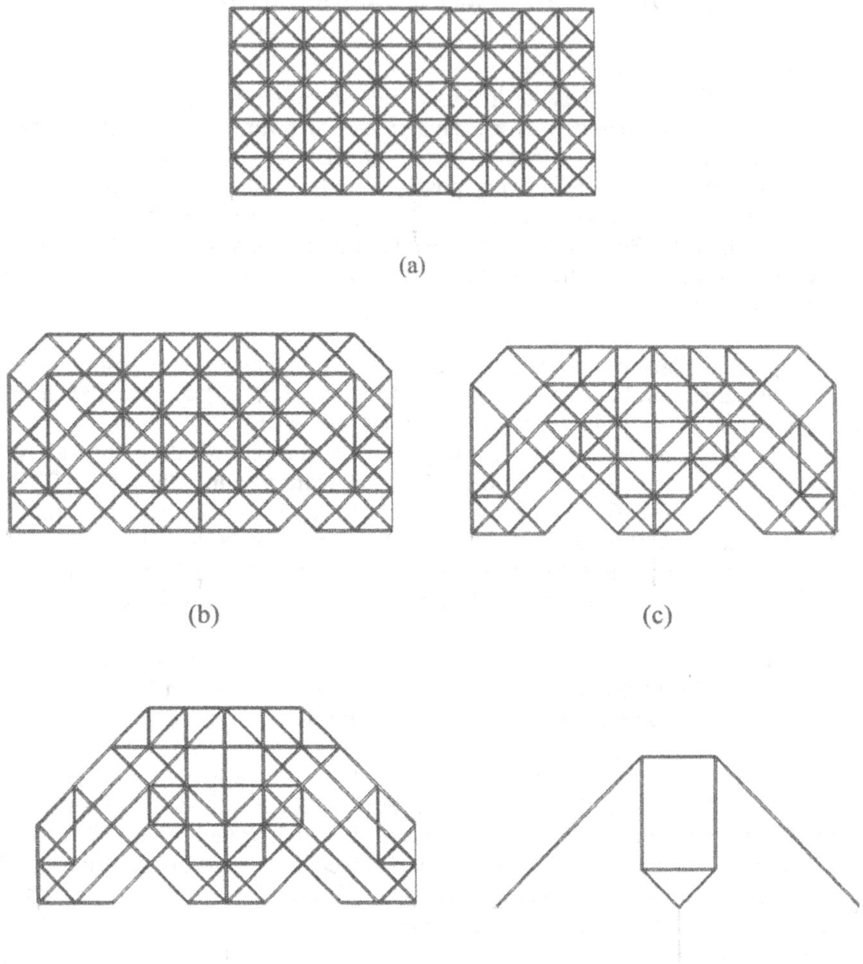

(a)

(b)

(c)

(d)

(e)

Fig. 7.9. Starting layout and evolving structure for topology optimization of a pin jointed frame.

The second illustrative example is that of a 3D cube supported at the bottom corners with a point load at the centre on the top. The base structure consists of 64 smaller cubes with all diagonals in use. A quarter of this base structure plus an interim and the final optimal topology are shown in Figure 7.10. This structure is optimized under a displacement constraint (Chu *et al.* 1996).

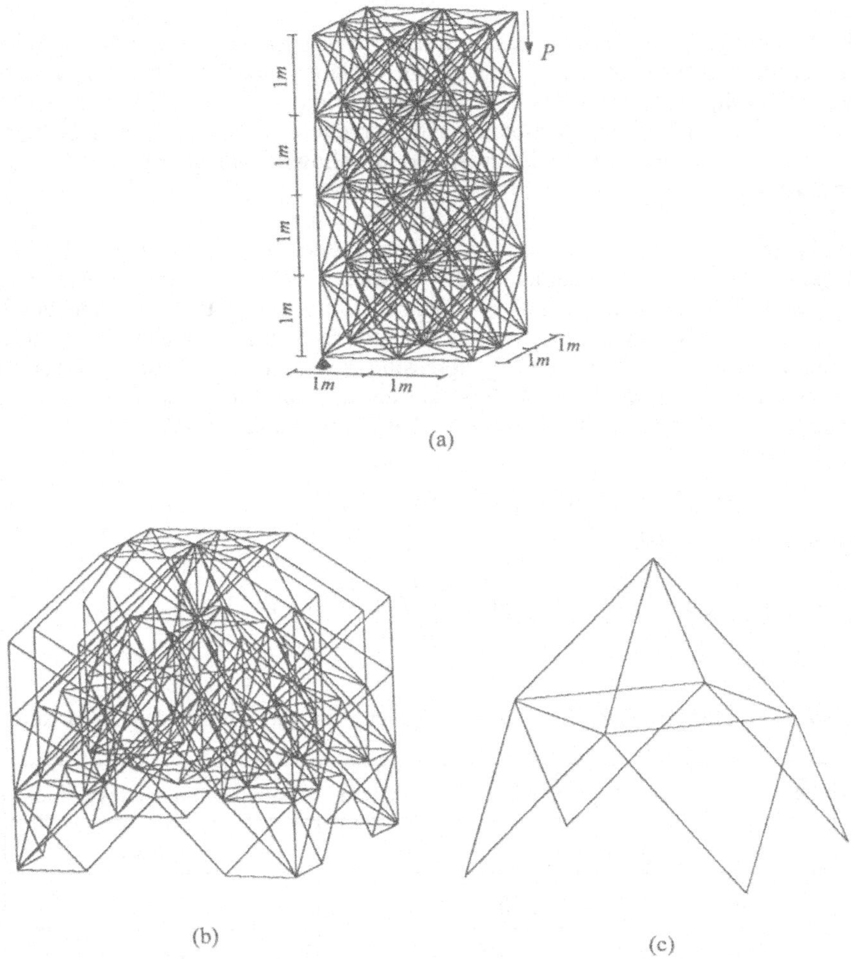

Fig. 7.10. ESO topology optimization for a 3D truss: (a) a quarter of the base structure; (b) an intermediate structure; (c) the optimal topology.

The topic of pin-jointed frame topology optimization has been covered extensively in the literature and these two examples presented here illustrate the directness of the ESO approach. Other examples of truss topology optimization by ESO have been reported by Dunn (1995) for fully stressed design and by Chu *et al.* (1996) for maximizing stiffness (or minimizing displacement).

7.6 Size Optimization for Beams and Rigid-Jointed Frames

The concept of ESO translates easily to beam type structures where there could be a range of desired outcomes. The design variables are some section properties such as width, depth or radius of a circular section. Frame members connecting nodes may be all of one section type for practical reasons or could be further subdivided to achieve a variable cross-section allowing for possibility of an even better result.

To illustrate the ESO process for rigid-jointed frames with bending as well as axial stresses and to act as a benchmark and validation example a cantilever beam with a tip load is first considered. For a length of unity and a unit load the bending moment is simply (1-x), where x is the distance from the fixed end. For a limiting maximum stress of 1 the following expressions can be obtained for width variation as a function of x when the depth h is constant at 1 and for the depth variation as a function of x when the width w is constant at 1. For width variation:

$$w_x = 6(1-x), \quad V = \int_0^1 6(1-x)dx = 3; \tag{7.2}$$

and for depth variation:

$$h_x = \sqrt{6(1-x)}, \quad V = \int_0^1 \sqrt{6(1-x)}dx = \frac{2\sqrt{6}}{3} = 1.633. \tag{7.3}$$

Table 7.7. ESO results for the optimal cantilever beam with width and depth variations.

No. of elements	Volume with width variation	Volume with depth variation
4	3.750	1.882
10	3.300	1.747
20	3.150	1.693
30	3.100	1.672
40	3.075	1.661

Table 7.7 gives the ESO results of the two options for size variation in the cantilever beam. These results and their convergence characteristics are totally consistent with the ESO logic. The decision to increase or decrease the width is based on the stress at the end of an element that has the highest stress. For a cantilever structure this is always the end nearest the wall. Consequently the width converged upon is that at the end nearer the wall. A simple calculation will show

that the constant depth beam elements can only have this thicker width and the error is the triangle of volume added by not being allowed to have a linearly varying width as in the theoretical optimum. The error is therefore the volume of all these little triangles given by one half times the maximum stress (in this case, 6) divided by the number of elements. For example with 20 elements the error is $0.5\times6/20 = 0.15$.

The results also indicate that the ESO process can be applied to rigid-jointed frames since each member of a rigid-jointed frame is in fact a beam.

The next example is a three storey 2D rigid-jointed building frame. Optimization is conducted with depth variation only. Two starting configurations generating three optimization models are proposed, designated as Model 1 to Model 3. Models 1 and 2 start from the shape of Figure 7.11(a). Model 1 uses 12 beam elements and Model 2 starts with 24 beam elements, being Model 1 with each beam element divided into two. Model 3 is Model 2 with 12 more beam elements providing some internal bracing. Figure 7.11 also shows the built-in freedom conditions at the ground, the mid-span downward loads of 20 kN on the cross beams and the sway loads of 10kN at the cross beam ends on the right hand side of the structures. All beam elements have a Young's modulus of 210 GPa, a fixed width of 0.05 m and an initial depth of 1.0 m. The stress limit is 3 MPa. The frame is 15 m high by 10 m wide.

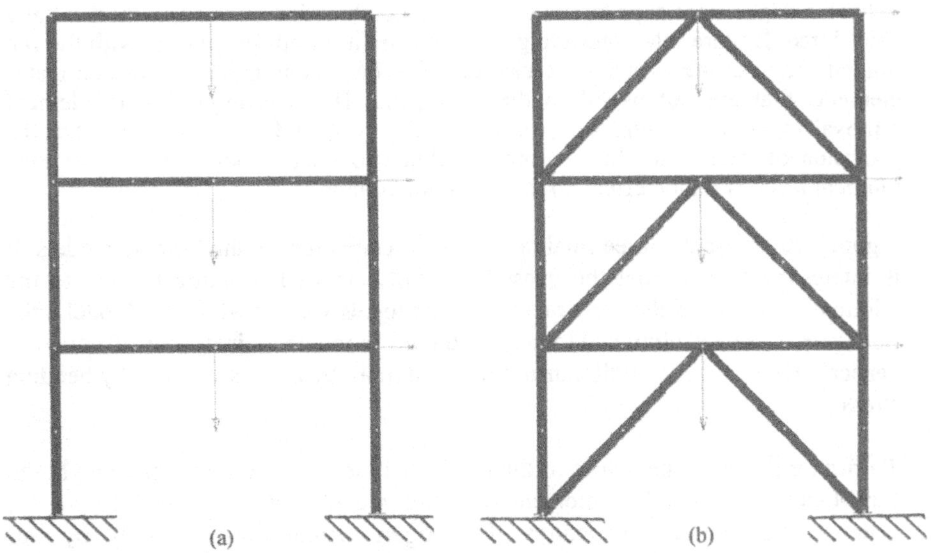

Fig. 7.11. Two starting structures for a three level rigid frame optimization.

Table 7.8 below gives details of the final optimal structures for different depth increments per iteration for the three frame models.

Table 7.8. Optimum volumes for three storey rigid-jointed frame for the three models with different depth increments.

Model	Depth increment (m)	Optimum volume (m^3)
1	0.002	2.86960
1	0.0002	2.86063
2	0.00025	2.55075
2	0.0002	2.55090
3	0.00125	0.50157
3	0.000125	0.50092

What can be deduced from Table 7.8 is that as with the cantilever the more elements used the lighter the final structure, thereby allowing each frame member to have a depth more appropriate to the local combination of bending moment and axial force. If more interconnecting members are allowed for then, as with the pin jointed frames, the structure seeks out the best combination and eliminates elements that are not useful to the load path. This is comparable to element removal in the continuum form of ESO. Also it should be pointed out that the provision of load paths that are more inclined to sustain axial force rather than moment has a dramatic effect on the optimum volume.

Figure 7.12 shows the three final optimum structures for the three frame models. It is interesting to note that the quest for a fully stressed structure has led to the elimination of part of the cross members for Models 1 and 2 whilst for Model 3 the ESO procedure has eliminated many of the cross members in favour of the more vertical members which will transmit the load more by axial stress than by bending stress.

To further illustrate the nature of the result that the ESO technique has sought out for Model 3, the bending moment diagrams are plotted in Figure 7.13 for the original and optimum structures. This figure demonstrates strongly by this graphical output that apart from the bottom right beam members there is very little bending in the rest of the structure. This observation is, in retrospect, what would be expected of an optimization routine for rigid-jointed structures with stress constraints, namely that, the algorithm seeks out a structure closest to the pin-jointed structure that would have the least volume, thereby eliminating the inefficiency of having load transmitted by bending rather than axial action.

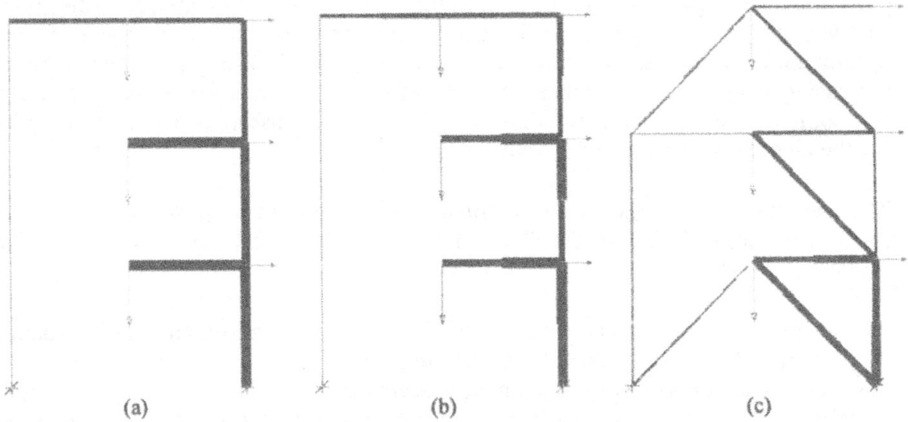

Fig. 7.12. ESO results for three starting models for the rigid-jointed frame. (Depth of members are indicative, but not to scale)

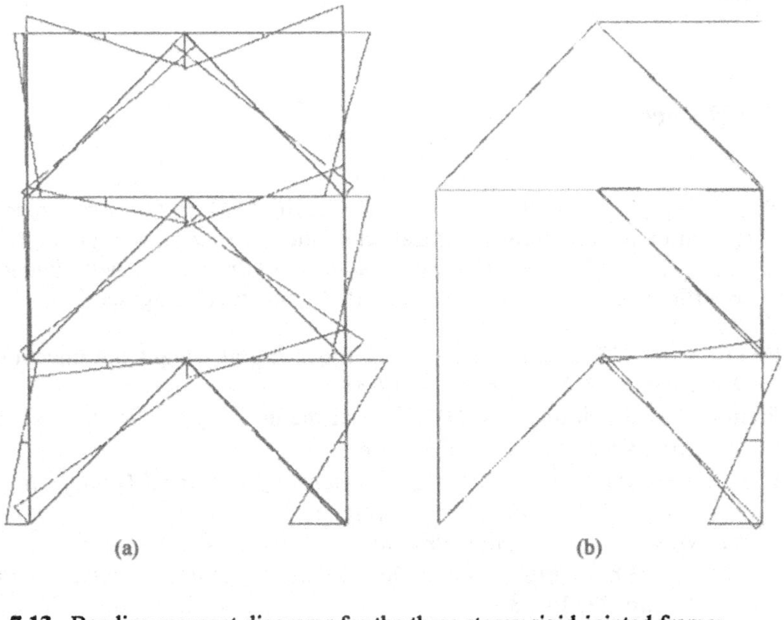

Fig. 7.13. Bending moment diagrams for the three storey rigid-jointed frame: (a) for the original structure; (b) for the optimal structure.

7.7 Conclusion

This chapter has demonstrated how the ESO concept can be simply applied to both pin-jointed and rigid-jointed frames in 2D and 3D. The computer code to do this is simple and robust and can be easily structured to solve much larger range of frame optimization problems. For example for multiple load cases the decision about increasing or decreasing depends on the member stress levels for all the load cases. The logic of the ESO process is unchanged, there are just more cases to consider for the element size change decision.

The most interesting conclusion that the authors would like to draw attention is that for a rigid-jointed frame the ESO process evolves towards a structure that minimizes bending.

When frame optimization is being considered there is the argument that in practice the material to be used comes from the supplier in fixed sizes and therefore it is unnecessary to conduct optimization with beams and frame members as continuous variables. Whilst it is true that in general there could be a restricted range of sizes available, there are two strong arguments in favour of being able to treat beam geometry as a continuous variable. Firstly the application may warrant obtaining the optimum volume regardless of fabrication details. Secondly having the absolute best structural size enables the designer to step back to the closest size available and know how far this compromise is away from being an absolute optimum.

7.8 References

Chu, D.N., Xie, Y.M., Steven, G.P. and Querin, O.M. (1996) Truss topology by an evolutionary structural optimization method, *Struct. Optim.* (submitted).

Dunn, E.L. (1995) *Evolutionary Structural Optimization with Beam Elements*, Fourth Year Thesis, Department of Aeronautical Engineering, University of Sydney.

Fleury, C. (1979) A unified approach to structural weight optimization. *Comput. Methods Appl. Mech. Eng.* **20**, 17-38.

Fleury, C. and Schmit, L.A. (1980) Dual methods and approximation concepts in structural synthesis. *NASA CR-3226*.

Haftka, R.T. and Gürdal, Z. (1992) *Elements of Structural Optimization*, 3rd edn., Kluwer Academic Publishers, Dordrecht.

Hemp, W.S. (1973) *Optimum Structures*, Clarendon Press, Oxford.

Kirsch, U. (1989) Optimal topologies of truss structures. *Comput. Methods Appl. Mech. Eng.* **72**, 15-28.

Kirsch, U. (1990) On the relationship between optimum structural topologies and geometries. *Struct. Optim.* **2**, 39-45.

Kirsch, U. (1993) Fundamental properties of optimal topologies. In: M.P. Bendsøe and C.A. Mota Soares (eds.), *Topology Design of Structures*, Kluwer Academic Publishers, Dordrecht, 3-18.

Michell, A.G.M. (1904) The limits of economy of material in frame-structures. *Phil. Mag.* **8**, 589-597.

Pederson, P. (1993) Topology optimization of three-dimensional trusses. In: M.P. Bendsøe and C.A. Mota Soares (eds.), *Topology Design of Structures*, Kluwer Academic Publishers, Dordrecht, 19-30.

Rozvany, G.I.N. (1989) *Structural Design via Optimality Criteria*, Kluwer Academic Publishers, Dordrecht.

Rozvany, G.I.N., Zhou, M, Birker, T. and Lewinski, T. (1994) Discretized methods for topology optimization. In: W. Gutkowski and J. Bauer (eds.), *Discrete Structural Optimization* (Proc. IUTAM Symposium, Zakopane, Poland, 1993), Springer-Verlag, Berlin, 135-147.

Rozvany, G.I.N., Bendsøe, M.P. and Kirsch, U. (1995) Layout optimization of structures. *Appl. Mech. Rev.* **48**, 41-119.

Sankaranarayanan, S., Haftka, R.T. and Kapania, R.K. (1993) Truss topology optimization with stress and displacement constraints. In: M.P. Bendsøe and C.A. Mota Soares (eds.), *Topology Design of Structures*, Kluwer Academic Publishers, Dordrecht, 71-78.

Topping, B.H.V. (1993) Topology design of discrete structures. In: M.P. Bendsøe and C.A. Mota Soares (eds.), *Topology Design of Structures*, Kluwer Academic Publishers, Dordrecht, 517-534.

7.9 Appendix

This appendix contains a sample code that can be used to undertake the size optimization examples of this chapter. The same source code can be found in the computer diskette, which is included with this book. The code has internal data set up and many of the more advanced optimization features are not present. The particular code given has the data set up to do rigid-jointed 2D ESO size optimization with a 35 bar ground structure similar in layout to that of Figure 7.5(b). The code is in standard FORTRAN and should compile in any computer. It is a simple task for the analyst to amend it to suit the desired design targets and variables.

```
program ESOFRAME
C****************************************C
C  EVOLUTIONARY STRUCTURAL OPTIMIZATION  C
C       Y.M. XIE   and  G.P. STEVEN      C
C****************************************C
C  Fortran source code to go with Chapter 7
C  ESO for Pin- and Rigid-Jointed Frames
C
C  Sample code does size optimization for rigid-jointed 2D frames
C  with the beam depth as the local design variable
C
C  The data in this file is the 2D rigid-jointed frame equivalent of
C  the 35 beam ground structure of Figure 7.5(b).
C
C  December 1996
C*************************C
C  VARIABLE DECLARATIONS  C
C*************************C
        implicit double precision (a-h,o-z)
        real*8 xy(50,2),forc(50,3),xl(2,3),dxy(50,3),
        real*8 u(6),ul(6),es(6,6)
        real*8 depth(66),strbem(66),SG(200,200),RHSG(200),disp(200)
        integer icon(66,2),istat(50,3),im(6)
        common/stiff2/SG,RHSG
C********************C
C  DATA DEFINITIONS  C
C********************C
        np=10                               ! number of points
        ne=35                               ! number of elements
        itlim=2000                          ! number of iterations
        depth0=0.5D00                       ! initial beam depth
        width=1.0D00                        ! fixed beam width
        ddepth=depth0*0.001D00              ! depth increment
        ymod=1.0d00                         ! Young's modulus
        sigtar=1.0d00                       ! target stress
C*************************C
C  NODE (X,Y) DATA INPUT  C
C*************************C
        xy(1,1)=0.0d0
        xy(1,2)=0.0d0
        xy(2,1)=-10.0d0
        xy(2,2)=0.0d0
        xy(3,1)=-8.660250d0
        xy(3,2)=5.0d0
        xy(4,1)=-5.0d0
        xy(4,2)=8.66025d0
        xy(5,1)=0.0d0
        xy(5,2)=10.0d0
        xy(6,1)=5.0d0
        xy(6,2)=8.66025d0
        xy(7,1)=8.66025d0
        xy(7,2)=5.0d0
        xy(8,1)=10.0d0
        xy(8,2)=0.0d0
```

```
      xy(9,1)=-10.0d0
      xy(9,2)=10.0d0
      xy(10,1)=10.0d0
      xy(10,2)=10.0d0
C*****************************************************************C
C NODE KINEMATIC STATUS FOR ROLLER END SUPPORTS: 0=FREE and 1=FIXED C
C*****************************************************************C
      do 20 i=1,np
            istat(i,1)=0
            istat(i,2)=0
20          istat(i,3)=0
      istat(2,1)=1
      istat(2,2)=1
      istat(8,2)=1
C******************C
C  BEAM CONNECTIONS  C
C******************C
      icon(1,1)=1
      icon(1,2)=2
      icon(2,1)=1
      icon(2,2)=3
      icon(3,1)=1
      icon(3,2)=4
      icon(4,1)=1
      icon(4,2)=5
      icon(5,1)=1
      icon(5,2)=6
      icon(6,1)=1
      icon(6,2)=7
      icon(7,1)=1
      icon(7,2)=8
      icon(8,1)=2
      icon(8,2)=3
      icon(9,1)=3
      icon(9,2)=4
      icon(10,1)=4
      icon(10,2)=5
      icon(11,1)=5
      icon(11,2)=6
      icon(12,1)=6
      icon(12,2)=7
      icon(13,1)=7
      icon(13,2)=8
      icon(14,1)=2
      icon(14,2)=4
      icon(15,1)=2
      icon(15,2)=5
      icon(16,1)=2
      icon(16,2)=6
      icon(17,1)=2
      icon(17,2)=7
      icon(18,1)=3
      icon(18,2)=5
      icon(19,1)=3
```

```
        icon(19,2)=6
        icon(20,1)=3
        icon(20,2)=7
        icon(21,1)=3
        icon(21,2)=8
        icon(22,1)=4
        icon(22,2)=6
        icon(23,1)=4
        icon(23,2)=7
        icon(24,1)=4
        icon(24,2)=8
        icon(25,1)=5
        icon(25,2)=7
        icon(26,1)=5
        icon(26,2)=8
        icon(27,1)=6
        icon(27,2)=8
        icon(28,1)=9
        icon(28,2)=2
        icon(29,1)=9
        icon(29,2)=3
        icon(30,1)=9
        icon(30,2)=4
        icon(31,1)=9
        icon(31,2)=5
        icon(32,1)=10
        icon(32,2)=8
        icon(33,1)=10
        icon(33,2)=7
        icon(34,1)=10
        icon(34,2)=6
        icon(35,1)=10
        icon(35,2)=5
C***********************************************C
C  FORCE INPUT - POINT DOWNWARD LOAD AT CENTRE  C
C***********************************************C
        do 30   i=1,np
                do 30 j=1,3
                     forc(i,j)=0.0D00
30      continue
        forc(1,2)=-1.0d00
C***************************************C
C  LOOP TO SORT OUT ROW NUMBERS IN [K]  C
C***************************************C
        nc=0
        ic=0
        do 40 i=1,np
                do 42 j=1,3
                        if (istat(i,j).gt.0) then
                                nc=nc-1
                                istat(i,j)=nc
                        endif
                        if (istat(i,j).eq.0) then
                                ic=ic+1
```

```
                                  istat(i,j)=ic
                        endif
42              continue
40      continue
        write(*,*) 'number of rows in global stiffness =',ic
C************************************************C
C  ALLOCATE ROWS TO NODAL DOF AND EVALUATE BANDWIDTH  C
C************************************************C
        mmm=0
        write(*,*) '========================================'
        write(*,*) '2D RIGID-JOINTED FRAME SIZE OPTIMIZATION'
        write(*,*) '========================================'
        write(*,*)
        write(*,*) 'D A T A '
        write(*,*) '-------'
        write(*,*) 'number of nodes = ', np,'   number of bars = ',ne
        write(*,*)
        write(*,*) 'iteration limit = ',itlim
        write(*,*)
        write(*,*) 'bar width = ',width
        write(*,*) 'initial bar depth =',depth0
        write(*,*) ' width increment =',ddepth
C*****************************************************C
C  ITERATION LOOP FOR ESO PROCESS WITH BEAM DEPTH CHANGING  C
C*****************************************************C
        iter=0
        do 280 i9=1,ne
                depth(i9)=depth0
280     continue
300     continue
C**********************************************C
C  ZERO OUT STIFFNESS MATRIX AND RHS LOAD VECTOR  C
C**********************************************C
        do 48 i=1,ic
                RHSG(i)=0.0D00
                do 48 j=1,ic
                        SG(i,j)=0.0D00
48      continue
        iter=iter+1
C************************C
C  ELEMENT ASSEMBLY LOOP  C
C************************C
        do 50 i=1,ne
                xl(1,1)=xy(icon(i,1),1)
                xl(1,2)=xy(icon(i,1),2)
                xl(2,1)=xy(icon(i,2),1)
                xl(2,2)=xy(icon(i,2),2)
                area=width*depth(i)
                secmo=width*depth(i)**3/12
                call bemstif(xl,area,secmo,ymod,es)
                im(1)=istat(icon(i,1),1)
                im(2)=istat(icon(i,1),2)
                im(3)=istat(icon(i,1),3)
                im(4)=istat(icon(i,2),1)
```

```
                im(5)=istat(icon(i,2),2)
                im(6)=istat(icon(i,2),3)
                if (iter.eq.1) then
                        do 52 j=1,5
                                if (im(j).lt.0) goto 52
                                do 54 k=(j+1),6
                                        if (im(k).lt.0) goto 54
                                        igap=abs(im(j)-im(k))
                                        if (igap.gt.mmm) mmm=igap
54                              continue
52                      continue
                endif
                do 62 i9=1,6
                        ll=im(i9)
                        if (ll.lt.0) goto 62
                        do 64 j9=1,6
                                kk=im(j9)
                                if (kk.lt.0) goto 64
                                SG(ll,kk)=SG(ll,kk)+es(i9,j9)
64                      continue
62              continue
50      continue
C*******************************C
C  CONSTRUCTION OF LOAD VECTOR  C
C*******************************C
        do 70 i=1,np
                do 72 j=1,3
                        if (istat(i,j).lt.0) goto 72
                        ir = istat(i,j)
                        RHSG(ir)=forc(i,j)
72              continue
70      continue
C**************C
C  INVERT [K]  C
C**************C
        call cholinv1(sg,ic)
C******************************************C
C  SOLUTION IS [K]**(-1) TIMES LOAD VECTOR  C
C******************************************C
        do 120 i9=1,ic
                sum=0.0
                do 122 j9=1,ic
                        sum=sum+sg(i9,j9)*rhsg(j9)
122             continue
                disp(i9)=sum
120     continue
C**************************************************C
C  SORT OUT SOLUTION VECTOR TO NODAL DISPLACEMENTS  C
C**************************************************C
        do 90 i=1,np
                do 92 j=1,3
                        nostat=istat(i,j)
                        if (nostat.lt.0) then
                                dxy(i,j)=0.0d00
```

```
                                else
                                dxy(i,j)=disp(istat(i,j))
                        endif
92              continue
90      continue
        totvol=0.0d00
        devres=0.0d00
        rmsdev=0.0d00
        nbars=0
C***************************C
C  ELEMENT STRESS RECOVERY  C
C***************************C
        do 80 i=1,ne
                xl(1,1)=xy(icon(i,1),1)
                xl(1,2)=xy(icon(i,1),2)
                xl(2,1)=xy(icon(i,2),1)
                xl(2,2)=xy(icon(i,2),2)
                dx=xl(2,1)-xl(1,1)
                dy=xl(2,2)-xl(1,2)
                el=sqrt(dx*dx+dy*dy)
                u(1)=dxy(icon(i,1),1)
                u(2)=dxy(icon(i,1),2)
                u(3)=dxy(icon(i,1),3)
                u(4)=dxy(icon(i,2),1)
                u(5)=dxy(icon(i,2),2)
                u(6)=dxy(icon(i,2),3)
                co=dx/el
                si=dy/el
C**************************************C
C  DISPLACEMENTS IN LOCAL COORDINATES  C
C**************************************C
                ul(1)=co*u(1)+si*u(2)
                ul(2)=-si*u(1)+co*u(2)
                ul(3)=u(3)
                ul(4)=co*u(4)+si*u(5)
                ul(5)=-si*u(4)+co*u(5)
                ul(6)=u(6)
                axstr=(ul(4)-ul(1))/el
C********************************************C
C  FIND MAX BENDING STRESS AT END 1 OF BEAM  C
C********************************************C
                bems=(-6.0*ul(2)-4.0*ul(3)*el+6.0*ul(5)-2.0*ul(6)*el)
                bems=bems*depth(i)/(el*el*2)
                stres1=(axstr + bems)*ymod
                stres2=(axstr - bems)*ymod
                end1=stres1
                if (abs(stres2).gt.abs(stres1)) end1=stres2
C********************************************C
C  FIND MAX BENDING STRESS AT END 2 OF BEAM  C
C********************************************C
                bems=(+6.0*ul(2)+2.0*ul(3)*el-6.0*ul(5)+4.0*ul(6)*el)
                bems=bems*depth(i)/(el*el*2)
                stres1=(axstr + bems)*ymod
                stres2=(axstr - bems)*ymod
```

```
                end2=stres1
                if (abs(stres2).gt.abs(stres1)) end2=stres2
                strbem(i)=end1
                if (abs(end2).gt.abs(end1)) strbem(i)=end2
                vol=width*depth(i)*el
                totvol=totvol+vol
                if (abs(depth(i)).le.1.0e-5) then
                        strbem(i)=0.0
                        goto 80
                endif
C**********************************************C
C  CALCULATE STRESS RESERVE FACTOR AND DEVIATION  C
C**********************************************C
                res=abs(strbem(i))/sigtar
C++++++++++++++++++++++++++++++++++++++++++++++C
                devres=devres+abs(res-1.0)
                if (depth(i).gt.(ddepth/100.0)) then
                        rmsdev=rmsdev+devres*devres
                        nbars=nbars+1
                endif
C****************************************************************C
C IF DEVIATION > 1 INCREASE BEAM DEPTH OTHERWISE DECREASE          C
C****************************************************************C
                if (res.gt.1.0) then
                        depth(i)=depth(i)+ddepth
                        elseif (res.lt.1.0) then
                        depth(i)=depth(i)-ddepth
                endif
C****************************************C
C  IF DEPTH HEADING TO ZERO SET VERY SMALL  C
C****************************************C
                if (abs(depth(i)).le.(ddepth/100.0)) then
                        depth(i)=0.0000001D00
                endif
C************************************************C
C  IF SOMEHOW DEPTH HAS SNUCK NEGATIVE SET VERY SMALL  C
C************************************************C
                if (depth(i).le.0.000D00) then
                        depth(i)=0.0000001D00
                endif
80      continue
C**********************C
C  END OF ESO ITERATION  C
C**********************C
        if (iter.gt.itlim) then
                write(*,*)
                write(*,*)
                write(*,*) 'iteration limit reached   --  final info'
                write(*,*)
                write(*,*) 'i vol rmsdev = ',iter,totvol,sqrt(rmsdev)
                write(*,*) '****************************'
                write(*,*) 'beam    stress       depth'
                write(*,*) '****************************'
                do 303 k9=1,ne
```

```
              write(*,304) k9,strbem(k9),depth(k9)

303           continue
304     format(I4,3X,2(E10.4,3X))
              stop
        endif
        goto 300
        end


C*********************************************************************C
C*********************************************************************C


        subroutine bemstif(xl,area,secmo,ymod,es)
C**************************************C
C  CALCULATES [K] FOR 2D BEAM ELEMENT  C
C**************************************C
        IMPLICIT DOUBLE PRECISION (A-H,O-Z)
        real*8 xl(2,2),es(6,6)
        dx=xl(2,1)-xl(1,1)
        dy=xl(2,2)-xl(1,2)
        el=sqrt(dx*dx+dy*dy)
        s=dy/el
        c=dx/el
        ek1=area*ymod/el
        ek2=12.0*ymod*secmo/(el**3)
        ek3=6.0*ymod*secmo/(el*el)
        ek4=4.0*ymod*secmo/el
        ek5=ek4/2.0
        es(1,1)=c*c*ek1+s*s*ek2
        es(1,2)=c*s*ek1-c*s*ek2
        es(1,3)=-s*ek3
        es(1,4)=-c*c*ek1-s*s*ek2
        es(1,5)=-c*s*ek1+c*s*ek2
        es(1,6)=-s*ek3
        es(2,2)=s*s*ek1+c*c*ek2
        es(2,3)=c*ek3
        es(2,4)=-c*s*ek1+c*s*ek2
        es(2,5)=-s*s*ek1-c*c*ek2
        es(2,6)=c*ek3
        es(3,3)=ek4
        es(3,4)=s*ek3
        es(3,5)=-c*ek3
        es(3,6)=ek5
        es(4,4)=es(1,1)
        es(4,5)=es(1,2)
        es(4,6)=-es(1,3)
        es(5,5)=es(2,2)
        es(5,6)=-es(2,3)
        es(6,6)=es(3,3)
        do 222 ii=1,6
        do 222 jj=ii,6
 222                es(jj,ii)=es(ii,jj)
```

```
        return
        end


C*****************************************************************C
C*****************************************************************C

      subroutine cholinv1(a,n)
C********************************************C
C  INVERTS [K] MATRIX USING CHOLESKY ALGORITHM  C
C********************************************C
      dimension a(200,200),c(200,200)
      IMPLICIT DOUBLE PRECISION (A-H,O-Z)
      ip=1
      do 1000 ii=2,n
1000  a(1,ii)=a(1,ii)/a(1,1)
1100  ip=ip+1
      ij=1
1005  ij=ij+1
      ir=0
      h=0.0
1010  ir=ir+1
      h=a(ip,ir)*a(ir,ij)+h
      if (ir.lt.(ij-1)) goto 1010
      a(ip,ij)=(a(ip,ij)-h)
      if (ip.gt.ij) goto 1005
1015  ij=ij+1
      ir=0
      h=0.0
1020  ir=ir+1
      h=a(ip,ir)*a(ir,ij)+h
      if (ir.lt.(ip-1)) goto 1020
      a(ip,ij)=(a(ip,ij)-h)/a(ip,ip)
      if (n.gt.ij) goto 1015
      if (n.gt.ip) goto 1100
      a(1,1)=1.0/a(1,1)
      do 1200 ii=2,n
      b=a(ii,ii)
      do 1300 ij=1, ii
      if (ii.ne.ij) goto  1030
      a(ij,ij)=1/a(ij,ij)
      goto 1300
1030  sum=0.0
      do 1040 il=ij, (ii-1)
1040  sum=a(ii,il)*a(il,ij)+sum
      a(ii,ij)=-sum/b
1300  continue
1200  continue
      do 1400 ii=1,n
      do 1500 ij=1,ii
1500  c(ii,ij)=a(ij,ii)
      c(ii,ii)=1.0
1400  continue
      c(1,1)=1.0/c(1,1)
```

```
      do 1600 ii=2,n
              b=c(ii,ii)
              do 1700 ij=1,ii
                      if (ii.ne.ij) goto  1750
                      c(ij,ij)=1.0/b
                      goto 1700
1750                  sum=0.0
                      do 1800 il=ij,(ii-1)
1800                  sum=c(ii,il)*c(il,ij)+sum
              c(ii,ij)=-sum/b
1700          continue
1600  continue
      do 1850 ii=1,n
              do 1860 ij=1,ii
                      sum=0.0
                      do 1870 il=ii,n
1870                  sum=c(il,ii)*a(il,ij)+sum
                      a(ii,ij)=sum
                      a(ij,ii)=sum
1860          continue
1850  continue
      return
      end
C*******************************************************************C
```

Chapter 8 ESO for Shape Optimization and the Reduction of Stress Concentrations

Ever since the formulation of the general rules and formulae of structural mechanics, the focus of most analyses has been to find the maximum stress present in a given structure. This maximum provided the basis of a design limit despite the obvious fact that, often, the bulk of the surface of the structure was at a much lower stress. This chapter demonstrates how the ESO technique can be used to provide structural shapes where the boundary is evolved on the basis of making the, as yet unknown, surface form as evenly stressed as possible.

8.1 Introduction

It is a well known maxim of structural designers that one should design the joints and other details first then design the rest of the structure. If a structure is going to fail then the germination of the failure is often the stress raised around some hole or corner or point of attachment or at some form of jointing, be it a weld or a mechanical fastener. It is these raised stresses and the presence of surface micro flaws that lead to crack initiation and eventually to static failure or fatigue.

The analysis of the stress at holes and corners and fasteners has occupied a special place in the history of the theory of elasticity and a large body of knowledge has built up to assist with the determination of stress concentrations at holes, corners and fasteners (Engineering Science Data Unit 1969; Peterson 1974).

It is traditional in design to attempt to eliminate stress concentrations. Techniques employed include reinforcing hole edges, reducing working loads, having no small radius curves, avoiding small radii in re-entrant corners. However, because structural shapes are often governed by the machines that make the parts, it often ends up that designs continue to have some stress raising geometric details. It is clearly logical to ask the question: what would be the best shape for a corner that minimizes the stress concentration? This is a typical shape optimization problem.

As has been demonstrated in the previous chapters, the ESO method caters for all three types of optimization, i.e. size, shape and topology optimization. This chapter presents the ESO method in its shape optimization guise. The articles by Haftka and Grandhi (1985) and Ding (1986) give excellent surveys of the state of shape optimization in the mid 1980's.

8.2 ESO for Shape Optimization

For shape optimization there is generally only a region of the structure that is of interest and not the whole structure. Thus only this region is available for the ESO process and the rest of the structure is deemed non-design. For shape optimization material is removed only from the structural boundaries. In ESO this is called *nibbling*.

The following is a brief description of ESO algorithm for shape optimization with stress based optimality constraints only. When stiffness is also a constraint the algorithm is different but only in detail at the time of removal.

1. Set up a dense mesh in the region to be optimized, the remaining non-design region can have a coarser mesh. If an outside profile is having its shape optimized then the model must have only that surface available for nibbling. If an internal surface is being considered then a small starting hole must be created first so that the internal surface is available for the nibbling process.

2. Apply the support conditions, loads and element properties. If there are several load cases these must be all included at this stage. If there are several sets of support environments then the same number of separate FEA models must be set up each with the load cases appropriate to that set of support conditions.

3. Undertake FEA analysis and extract element stresses for each load case in each support environment. For design elements compare the von Mises stress of each surface element with the maximum stress. Remove the element if the element stress satisfies condition (2.2) in Chapter 2 for all load cases and all support environments.

4. Repeat 3 till no elements are removed for the fixed Rejection Ratio (*RR*). When such a steady state is reached, increase *RR* by *ER*.

5. Repeat 3 and 4 till some specified value of *RR* has been reached or till some specified amount of material has been removed.

Typically, for shape optimization, the initial rejection ratio RR_0 has to be low say 1% and *ER* has also to be low say 1%. Often it may take 10 to 20 iterations at each *RR* before a steady state is reached.

Typically shape optimization has to do with the detail of a structure or structural component, for example the shape of a hole in a complex stress field or the shape of the spew of glue on an adhesively bonded joint. This form of optimization has as an objective not so much to reduce the weight of a structure as to reduce local stresses. It is these high local stresses that lead to fatigue failure. There is no point in having a structure whose topology has been optimized as in Chapters 2 and 3 if a sudden change of section or sharp corner is introduced that causes a 3 or 4 times stress concentration.

8.3 Examples

8.3.1 Fillet in Plane Stress under Tension

This has been chosen as the first example as it is a classical situation which has been much studied in the plane stress mode as well as the axisymmetric mode. When a bar in tension has to change its width a fillet with a certain radius is traditionally used to avoid the obvious singularity at the re-entrant corner. Item 69020 in Engineering Science Data Unit (1969), for example, has a whole sheet on this situation for various width changes and corner radii.

With the ESO process there are several aspects that can be studied. Firstly for a given neck length what profile of fillet will produce the minimum possible stress concentration. The neck length is the distance along the thinner portion that the fillet region is allowed to spread. In this study four fillet lengths are considered. These are namely 0.25, 0.5, 1.0 and 2.0 times the thicker width of the whole structural shape. Once the profile has been determined it is of interest to know what neck length provides the absolute minimum possible stress concentration.

The second aspect of interest is the level of stress concentration in the optimized fillet shape as compared to that by the traditional circular fillet profile using the same amount of material.

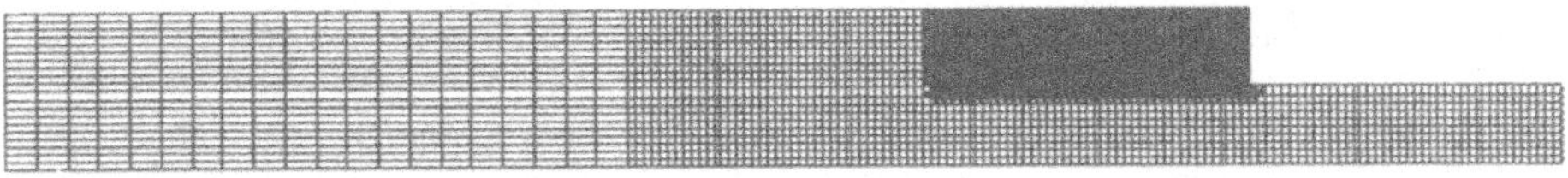

(a)

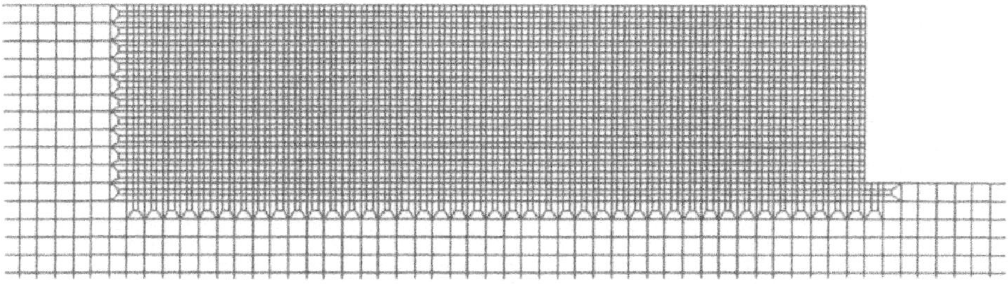

(b)

Fig. 8.1. Initial FEA half model for fillet with transition length equal to the thicker width: (a) the whole mesh (bottom and LHS are symmetry planes, and RHS has unit normal stress applied); (b) detail of the fillet region.

Figure 8.1 shows the initial mesh for the ESO process when the neck length is equal to the thicker width. Only half of the full structure needs to be modelled and the thicker portion is extended for the allowable neck length and divided into a fine mesh of four node elements which are all available for removal during the ESO process. All the other parts of the structure are non-design type elements and cannot be removed. Figure 8.2 shows eight pictures of close-ups on the design material as it evolves. Initially the top right portion of the region where one would expect the stress to be least is removed. As the evolution progresses the profile changes from the hump at low rejection ratios through an S shape to being concave. When rejection ratio $RR = 38\%$ the length of the fillet still fits inside the allowable neck length, but thereafter the space taken up by the fillet reduces as is seen in the picture corresponding to $RR = 40\%$. From $RR = 40\%$ onwards the fillet quickly shrinks into the corner.

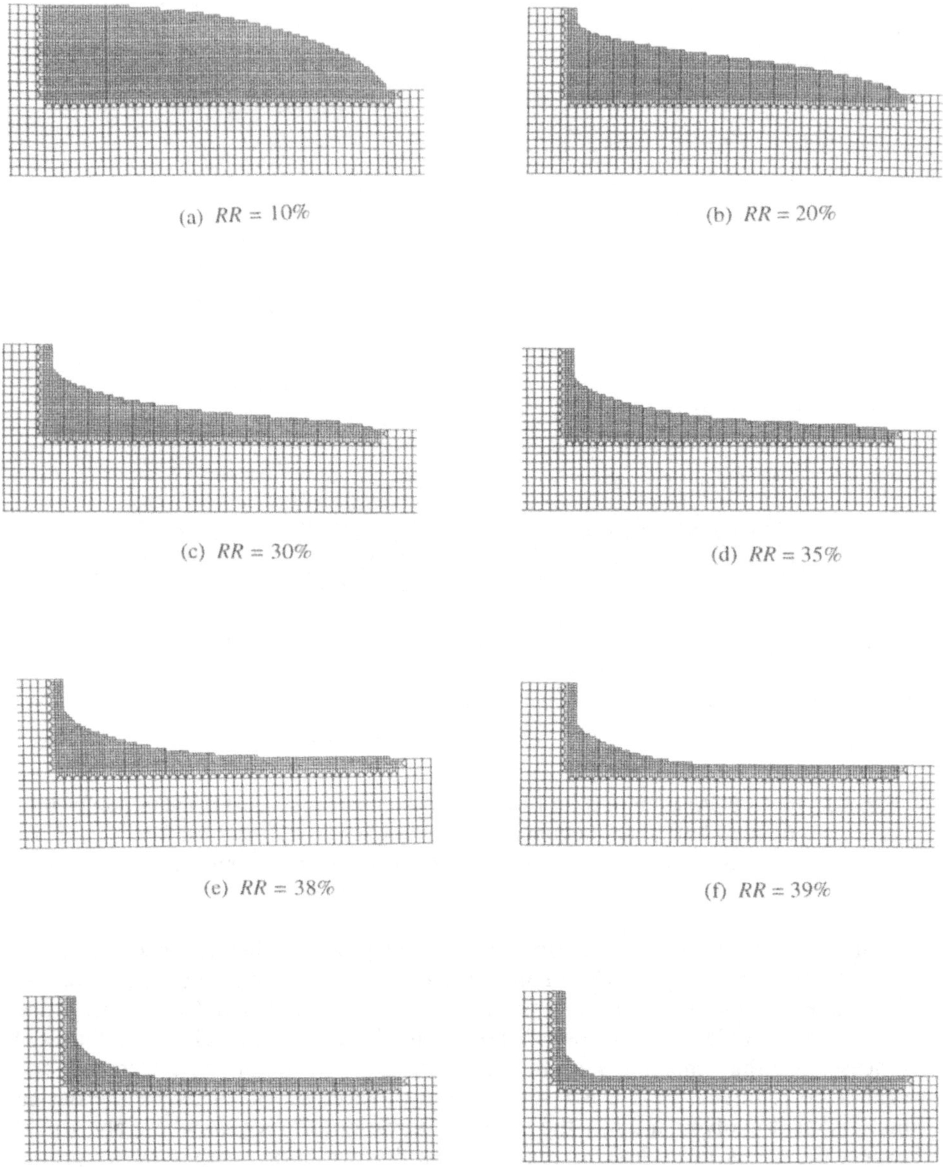

(a) $RR = 10\%$ (b) $RR = 20\%$ (c) $RR = 30\%$ (d) $RR = 35\%$ (e) $RR = 38\%$ (f) $RR = 39\%$ (g) $RR = 40\%$ (h) $RR = 50\%$

Fig. 8.2. Evolution of the fillet with transition length equal to thicker width.

It is interesting to note that the pictures in Figure 8.2 have been shown to several experienced stress analysts who were asked to select the profile which would give the smallest stress concentration ratio. All participants in this quiz selected the *RR* = 38% profile. This instinctive result corresponds exactly to the lowest value of maximum von Mises stress encountered at any of the ESO steady states. For all the evolution sequences for the four transition neck lengths the maximum von Mises stress at each steady state is plotted against steady state number in Figure 8.3. This clearly shows that for the neck length equal to the thicker width the maximum stress goes progressively down until *RR* = 38% and moves upwards steeply after that.

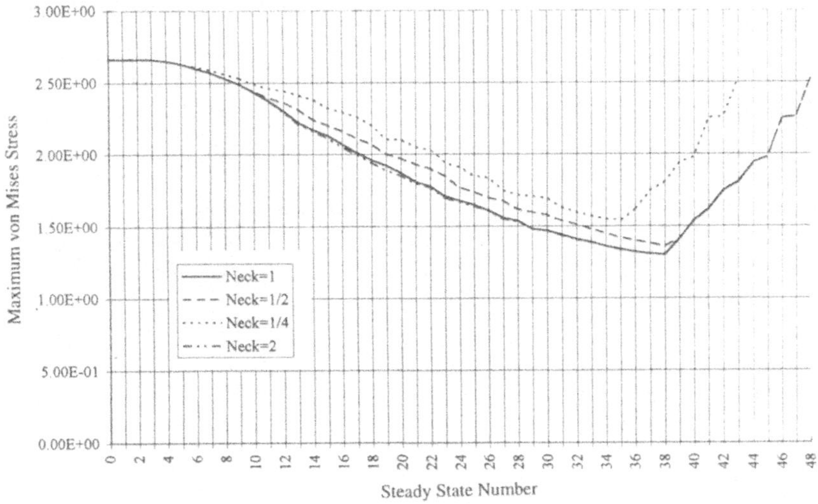

Fig. 8.3. Plots of maximum von Mises stress against steady state number for four neck lengths.

The corresponding evolution sequences for neck lengths equal to 1/4, 1/2 and 2 times the thicker width are shown in Figures 8.4, 8.5 and 8.6, respectively.

The second point of interest to the stress analyst is what improvement has been achieved by the ESO process. To answer this question consider the fillet when the neck length is equal to the thicker width. Figure 8.2 shows the *RR* = 38% optimum profile where the maximum stress reaches the lowest value. If the area occupied by the fillet were converted to fill a circular fillet shape then the maximum stress generated by the circular fillet can be established from Figure 65 in Peterson (1974). The ratio of the equivalent radius to the thinner width is about 0.5 and it is necessary to extrapolate Peterson's Figure 65 to give a stress concentration of

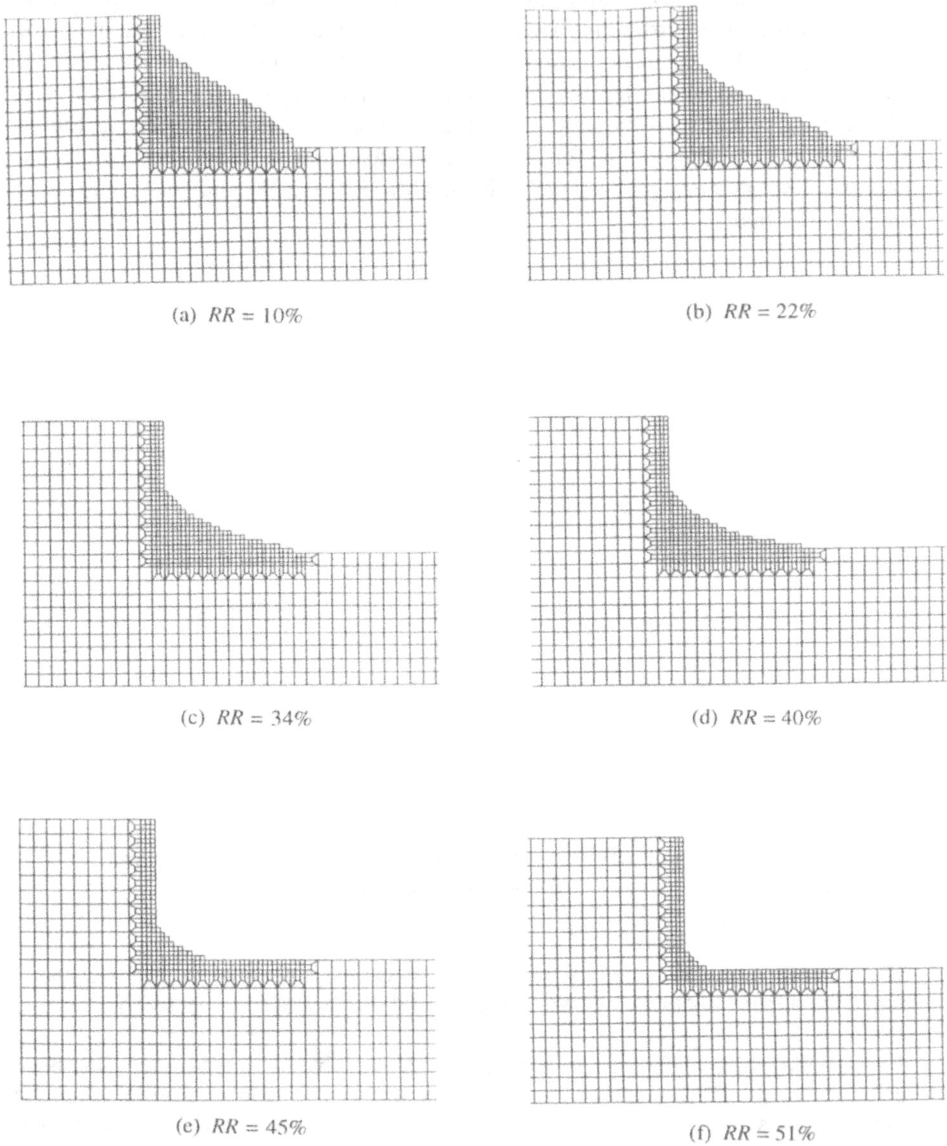

(a) $RR = 10\%$

(b) $RR = 22\%$

(c) $RR = 34\%$

(d) $RR = 40\%$

(e) $RR = 45\%$

(f) $RR = 51\%$

Fig. 8.4. Evolution of the fillet with transition length equal to a quarter thicker width.

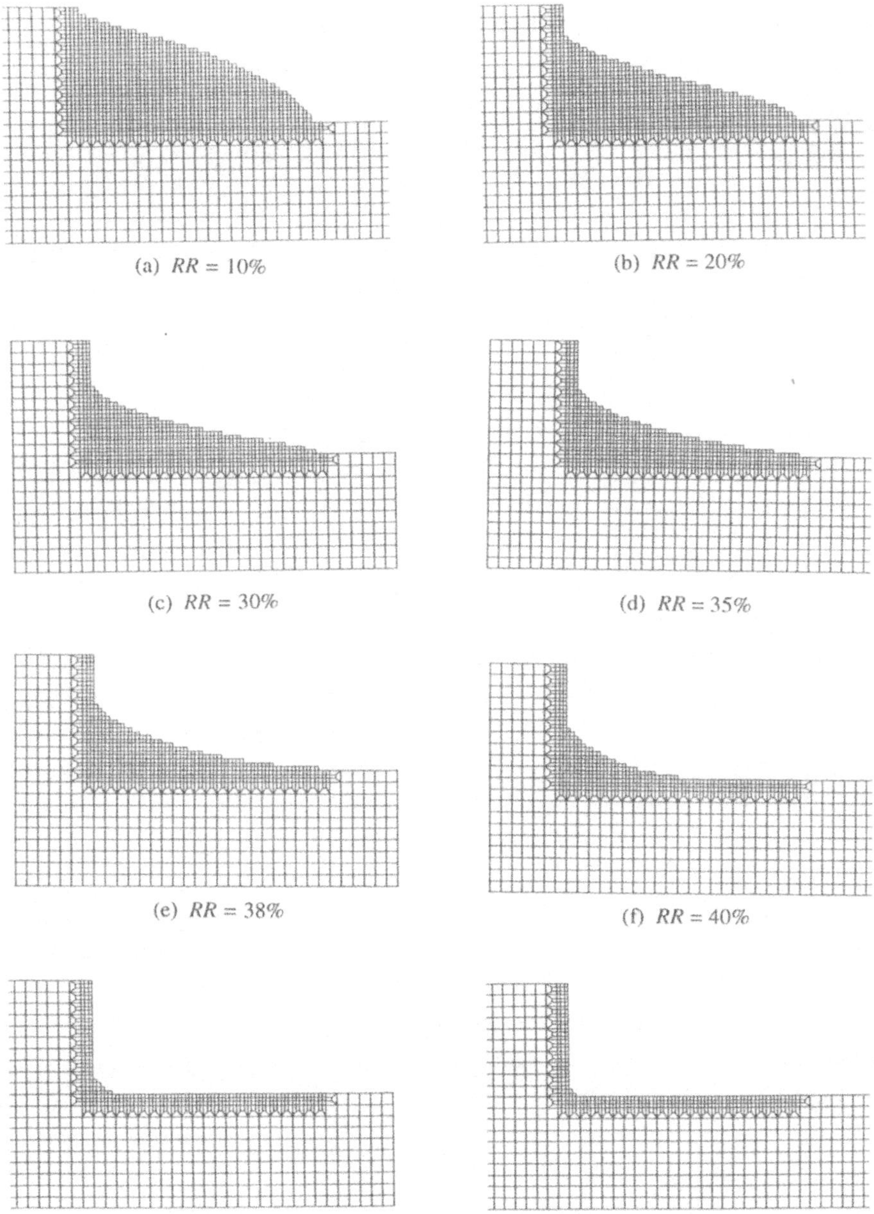

(a) $RR = 10\%$

(b) $RR = 20\%$

(c) $RR = 30\%$

(d) $RR = 35\%$

(e) $RR = 38\%$

(f) $RR = 40\%$

(g) $RR = 50\%$

(h) $RR = 60\%$

Fig. 8.5. Evolution of the fillet with transition length equal to half thicker width.

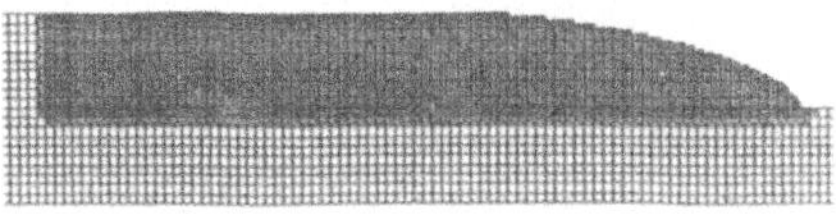

(a) $RR = 10\%$

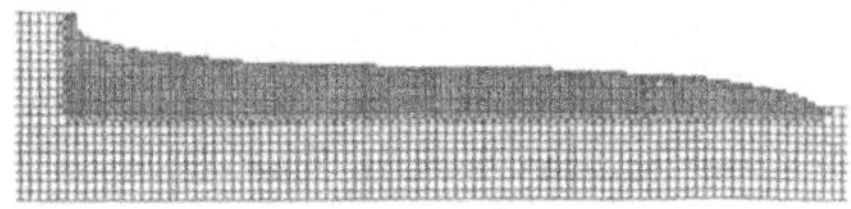

(b) $RR = 20\%$

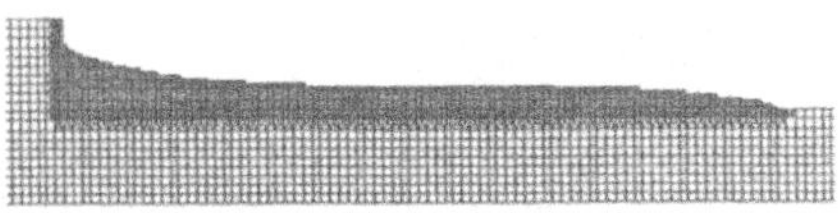

(c) $RR = 25\%$

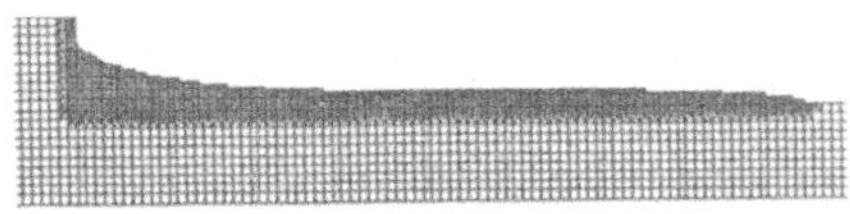

(d) $RR = 30\%$

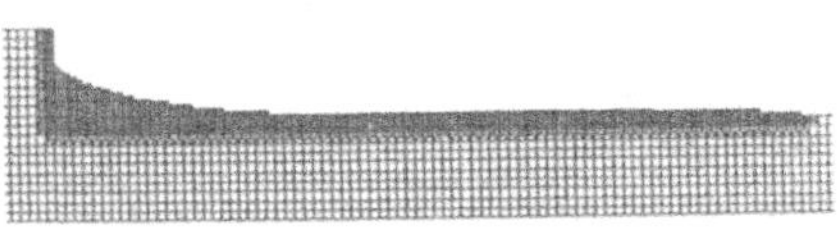

(e) $RR = 35\%$

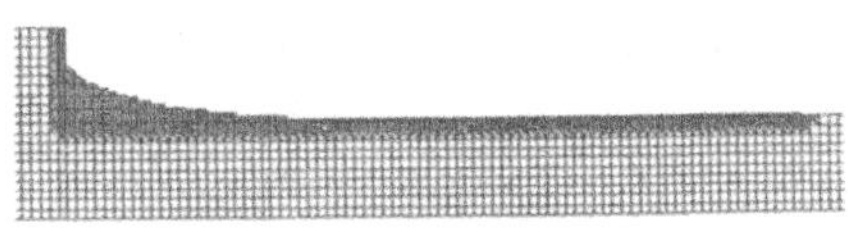

(f) $RR = 38\%$

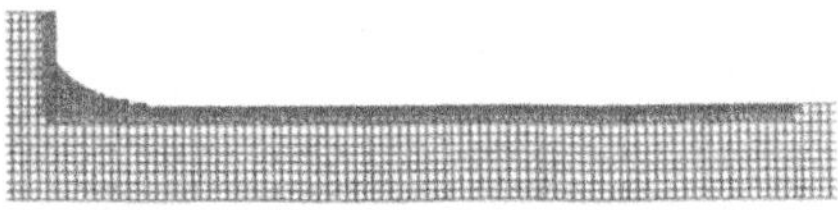

(g) $RR = 40\%$

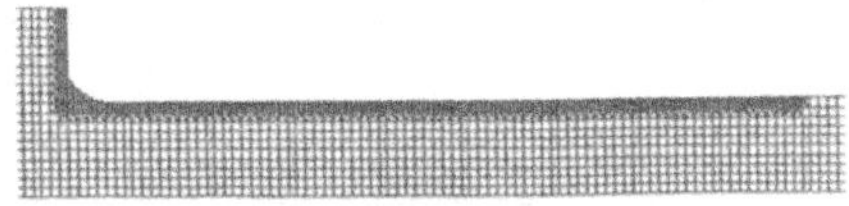

(h) $RR = 45\%$

Fig. 8.6. Evolution of the fillet with transition length equal to twice thicker width.

approximately 1.6. This is to be compared with 1.3 from the ESO procedure. In terms of the stress raise this means the simple circular arc fillet doubles the stress concentration. Similar increases are obtained for the other neck lengths except the shortest one where there is not enough axial length available for the ESO process to fully achieve an optimum solution.

Finally from this simple study there is a question about what length should be allowed to maximize the benefit from a fully developed fillet. By comparing the *RR* = 38% fillet shapes for neck length equal to one and two times the thicker width, as shown in Figures 8.5 and 8.6, it is seen that the fillet occupies about the same axial length of the bar for the *RR* at which the maximum stress reaches the lowest value and that this lowest value is the same for both neck lengths, as can be seen in Figure 8.3. In other words the ESO process has produced the same optimized profile for both cases and there was no need to have the neck length longer than the shorter of the two. But when the neck length is too short the minimum value of the maximum von Mises stress in Figure 8.3 increases.

In this example the thinner width has been fixed in the initial FEA models to be equal to half the thicker width. If the thinner width were of a different proportion of the thicker width, the optimum transition length and the minimum value of the maximum von Mises stress could have been different.

8.3.2 Cantilever with a Tip Load

One of the most common structures is a simple cantilever with a tip load. There are two aspects of the design optimization: one is to minimize the mass of the cantilever, and the other the maximum stress. The first is the classical question of the variation of the depth of the cantilever such that the maximum fibre stress is constant. Using the traditional bending theory the result is a quadratic variation in depth. The deficiency in this shape is that from elasticity theory it can be shown that there is a one over radius to the quarter power singularity at the re-entrant corners where the cantilever joins the wall. Thus the strategy of the ESO method for shape optimization would be to capture both the quadratic depth variation of the bulk of the cantilever plus determine what shape of fillet is needed in the corners to avoid the singularity.

Whilst a solid cantilever can never be a fully stressed design because of the linear variation of bending stress, the means of assessing the shape optimized design would be to have the outside profile geometrically coincident with the stress contour. This logic is validated by considering the inverse situation where contours start or finish on the outside profile, which indicates that there are regions of lower stress round the contour and that material is not distributed to its best.

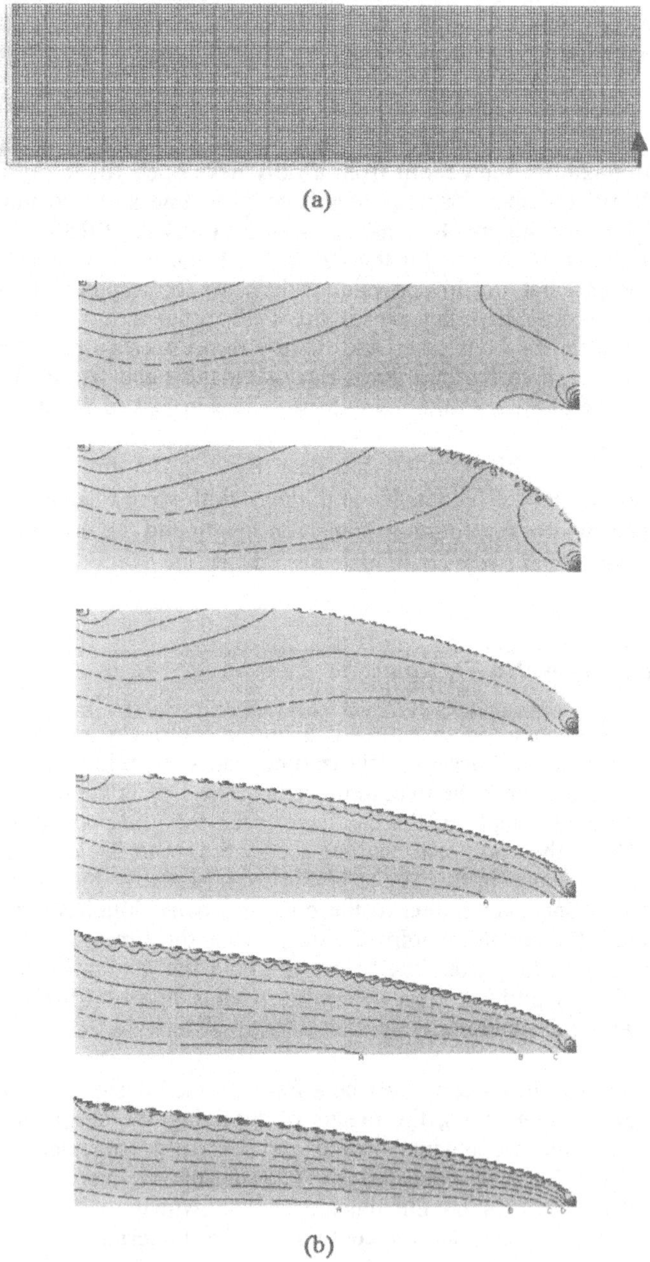

Fig. 8.7. Upper half of cantilever with a tip load: (a) FEA mesh; (b) shapes and stress contours during evolution.

Figure 8.7(a) shows the initial FEA mesh for this shape optimization scenario. Because the bulk of the cantilever is expected to evolve to the quadratic curve the mesh is not so refined there. Near the root there is a fine mesh. The evolutionary process will find the optimum shape which has the stress contour going full round the perimeter as well as the finite element mesh resolution will allow. If the evolutionary process is allowed to continue further all that happens is that scaled down identical forms emerge, each one being an optimum for the total depth of beam involved.

The rest of Figure 8.7 shows several steady states of the beam shape as it evolves. Figure 8.8 shows an optimum shape of cantilever with details at root corner and at tip.

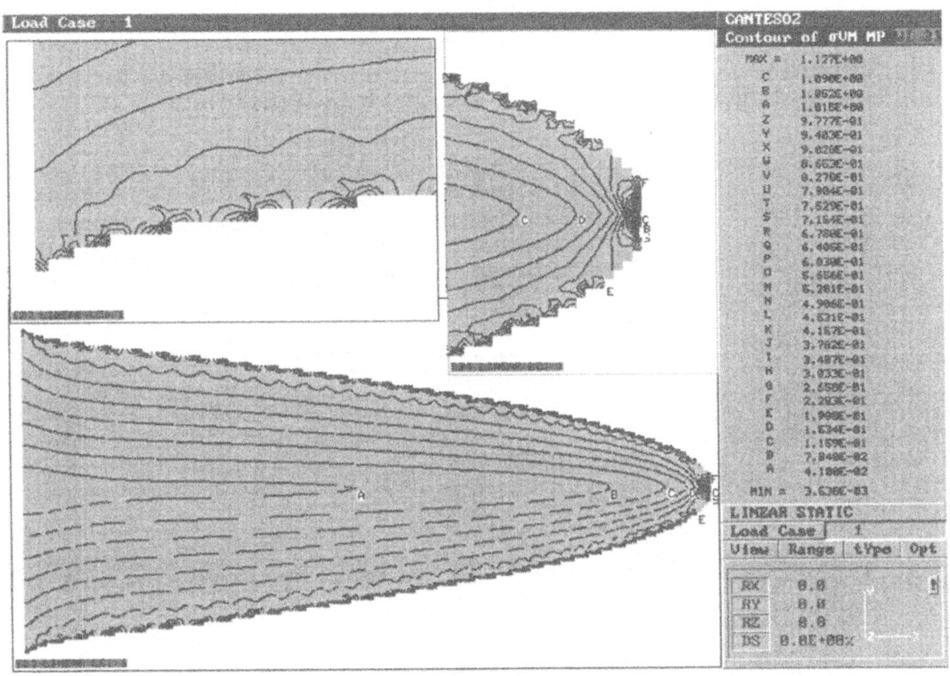

Fig. 8.8. Final optimized shape and details at root corner and at tip.

8.3.3 An Open Ended Spanner

The shape of an open ended spanner is a highly evolved one. Spanners have been in existence since the 16th century when they appeared as a complementary development to the screw thread cutting process. Many evolutionary changes have occurred to produce the drop forged tools we currently use. The loading on a spanner is simple to prescribe in terms of point contact with two ends of opposing flats on a nut, thus producing a torque.

Figure 8.9 shows the initial design domain, the non-design regions, the loading and freedom conditions for this study in shape optimization. Note that two equal but opposite load cases are needed to satisfy the bi-directional use of spanner. Figure 8.10 shows the evolutionary sequence for the spanner as it evolves towards the final shape shown Figure 8.10(e). Figure 8.10(f) is a spline fit through the final FEA mesh. This could be sent to a numerically controlled mill to produce the optimum shape. Such a rapid design and prototyping is likely to become an important part of product development.

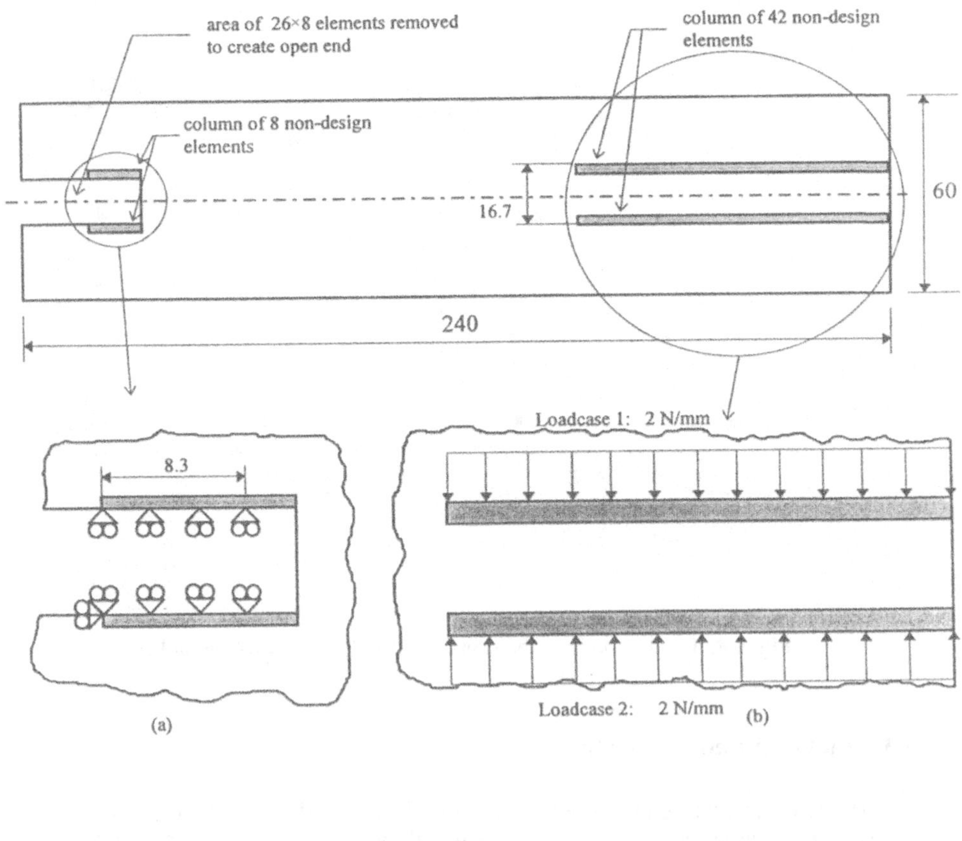

Fig. 8.9. Initial design space, loads and freedom conditions for spanner.

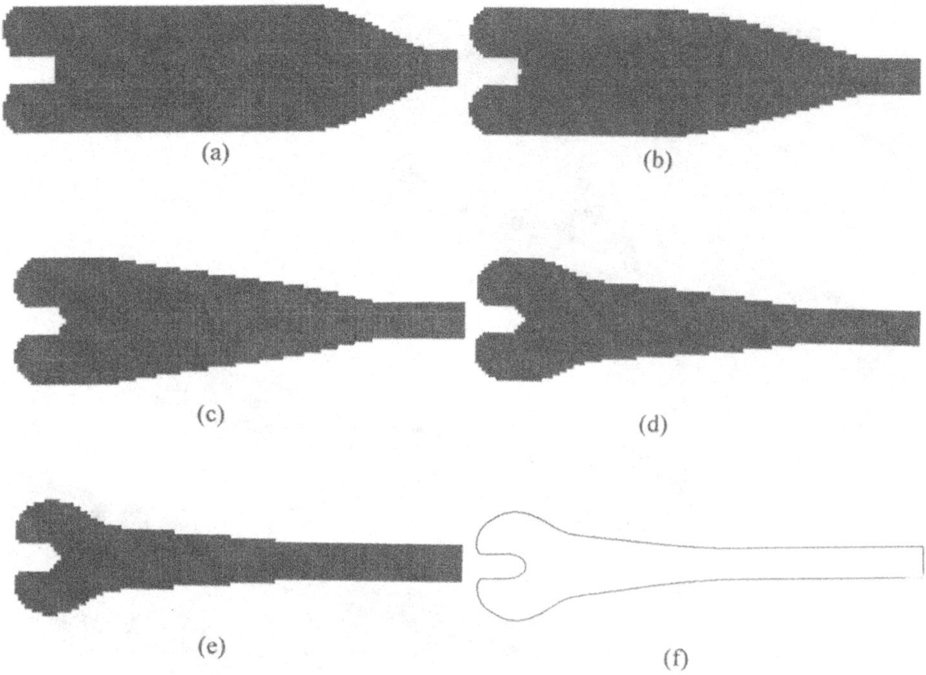

Fig. 8.10. Evolution of spanner shape and spline fit to final shape.

The shape in Figure 8.10(f) closely resembles what we have come to know as a spanner. This clearly demonstrates that the natural development of the modern spanner has been along the correct lines. As can be appreciated the nibbling form of ESO does not take into account the stress state within the continuum. It simply drives the surface towards being evenly stressed. Because of the bending involved in the action of a spanner there is very lowly stressed material in the interior. For this reason extensions of the ESO process can take the shape at the end of the 2D plane stress evolution and conduct further iterations whereby 3D bricks replace the plane stress elements and the lowly stressed bricks can have their thickness evolved downwards to increase their stress level thereby moving more towards a stress equalized shape. We call this process *morphing* to distinguish it from *nibbling*. Figure 8.11 shows the results of this morphing process for the spanner and again the connection between the evolved shape and the natural evolution of the real spanner is remarkable.

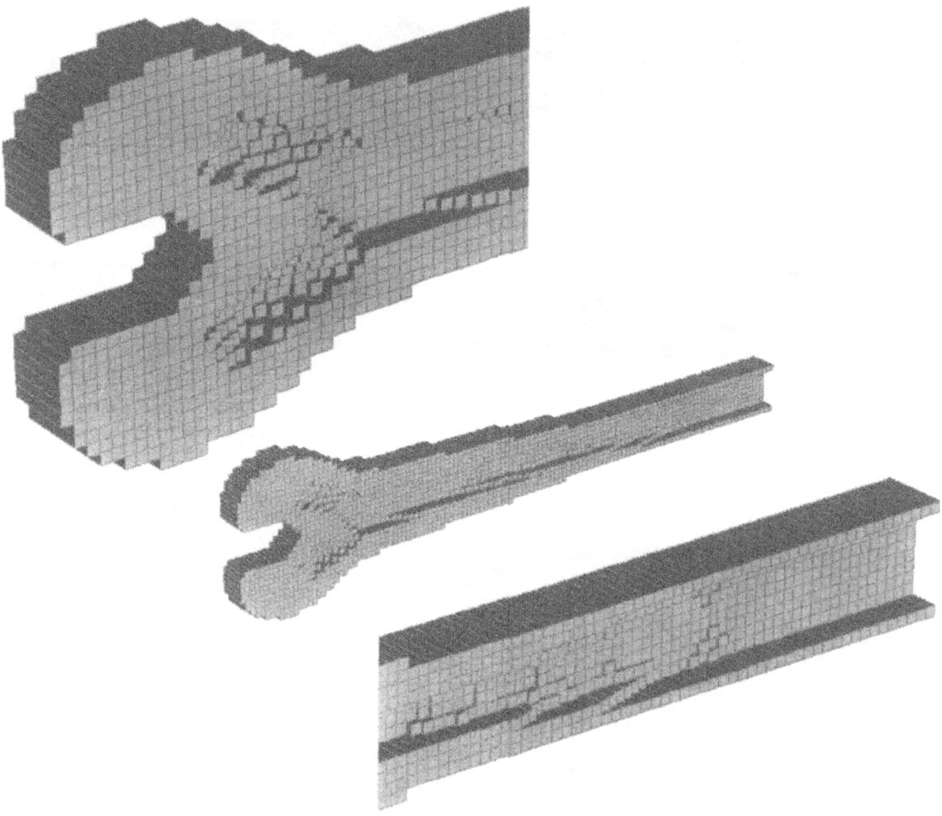

Fig. 8.11. Morphing of spanner to allow thinner interior region.

8.3.4 Optimum Shape of Holes

The stress concentrations caused by the cutting of a hole can significantly change the strength and fatigue life of the structure. Being able to find the shape of the hole that raises the maximum stress by the least amount is of great interest.

If as part of the design of a structure cutouts for the likes of windows are necessary then the shape of the cutout can be structurally important. For a 2D biaxial stress state, classical analysis gives the result of a hole shape that minimizes the maximum hole stress in an ellipse whose axes are in the direction of and in proportion to the ratio of the edge stresses.

When the cutout is near an edge, or in a composite material or near another cutout, the optimum shape of the hole needs to be found by shape optimization.

To use ESO for this type of problem, the design region needs to have the usual fine mesh. The elements around the outer boundary are made non-design and a small starting hole (or holes, if a multiple hole simulation is being undertaken) is made by removing one or a few elements. The software is set so that *nibbling* is allowed only at the surface of the hole.

Since the maximum stress is generally somewhere on the perimeter the rejection of lowly stressed elements has the now familiar effect of driving the surface towards being as evenly stressed as possible. Once this situation is reached further evolution simply increases the size of the hole.

As the hole size becomes a significant fraction (say greater than 10%) of the whole structural region, the boundaries begin to play a part and the optimum hole shape starts to change.

For interest we shall use an example of a laminated composite material of which an isotropic material is a special case. With an isotropic material an element on the surface of the hole would be removed if its von Mises stress is low. When the material is a laminated composite the Tsai-Hill number is used rather than the von Mises stress. The Tsai-Hill number for a single ply with material axes (l, t) is

$$TH = \left(\frac{\sigma_l}{X}\right)^2 + \left(\frac{\sigma_t}{Y}\right)^2 - \left(\frac{1}{X^2} + \frac{1}{Y^2}\right)\sigma_l\sigma_t + \left(\frac{\tau_{lt}}{S}\right), \tag{8.1}$$

in which X and Y are the appropriate tensile or compressive failure stresses in the two material axes and S is the shear failure stress.

By adopting a first-ply-failure philosophy (Tsai 1988), the ply with the maximum Tsai-Hill number within each plate is chosen as the limiting ply and the corresponding Tsai-Hill number as the limiting Tsai-Hill number, which is denoted TH^i for plate i. An element on the surface of a hole is removed if its limiting Tsai-Hill number is less than the rejection ratio RR times the average limiting Tsai-Hill number of all the plates in the FEA model. Mathematically this is expressed as

$$\text{Remove plate } i \text{ if} \quad TH^i < \frac{1}{N}\sum_{j=1}^{N} TH^j \times RR, \tag{8.2}$$

where N is the total number of plate elements. Typical carbon fibre/epoxy composite data is used as given in the following table.

Table 8.1. Material properties for carbon fibre/epoxy composite.

E_l	128.0 GPa
E_t	11.3 GPa
G_{lt}	6.0 GPa
ν_{lt}	0.3
Ply thickness	0.16 mm
Longitudinal tensile strength (*X*)	1.45 GPa
Longitudinal compressive strength (*X*)	1.25 GPa
Transverse tensile strength (*Y*)	52 MPa
Transverse compressive strength (*Y*)	100 MPa
Shear strength (*S*)	93 MPa

In this example a sheet of material is subjected to pure shear. The material is either a $[\pm 45/0/90]_s$ symmetric layup meaning the 8 plies are at +45, -45, 0, 90, 90, 0, -45, +45 to the principal axis of the sheet, or it is an isotropic material. Figure 8.12 shows the starting hole and the optimized holes for the two material types. A measure of the efficacy of the ESO process would be to compare the maximum Tsai-Hill number or von Mises stress for the optimum against the same number or stress if a circular hole of the same area were used. For the composite material the ESO optimum is 40% lower and for the isotropic material the ESO optimum is 37% lower. The fact the optimum shape is rectangular or even square for isotropic material is noted by Backlund and Isby (1988).

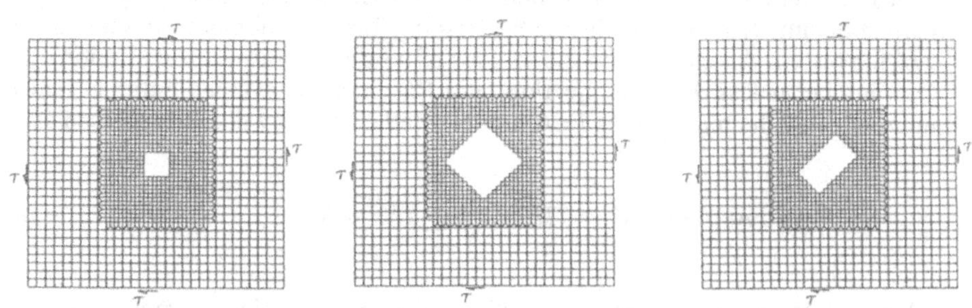

(a) (b) (c)

Fig. 8.12. Hole in a sheet of material subjected to pure shear: (a) starting hole; (b) optimum hole for isotropic material; (c) optimum hole for composite material.

Further details of cutout shape optimization using ESO for single and multiple holes can be found in Falzon (1996) and Falzon *et al.* (1996).

8.3.5 An Adhesive Joint

Joining two or more structural components often leads to stress concentrations and joints provide sources of structural failure. It is sometimes said that to undertake structural design on an airframe the analyst simply has to do the joints first and the rest of the structure will be adequate. This leads to a design hypothesis for a joint that its presence is neutral with respect to the stress state around it. In other words the joint does not raise the stress above the prevailing level. This statement is in accordance with the way the ESO process works for shape optimization.

When the analyst has no concept of what the best shape is going to be, conservatism leads to the choice of more rather than less material. Starting from such a decision the ESO process will remove material till the final evolved shape has no low stressed material at the surface and it is close to being an iso-stress line.

Only one form of the many options for an adhesively bonded joint is considered, namely a double strap butt joint where two separate sheets are butted together and straps are adhesively attached either side. Figure 8.13 shows the arrangement and the dimensions of the joint. The sheets and straps are aluminium with a Young's modulus of 70 GPa and Poisson's ratio of 0.33 and the adhesive film, which fills the gap between the sheets and straps, is epoxy with a Young's modulus of 4 GPa and Poisson's ratio of 0.38. Note that there is no adhesive between the ends of the sheets. Due to symmetry, only one quarter of the joint need to be analyzed. The ESO process allows material to be nibbled away from the outer surface of the strap only.

Figure 8.14 shows the evolving shapes of the strap as the rejection ratio *RR* increases. In each shape the stress contours have been plotted and although they cannot be readily interpreted what can be seen is that as the profile of the top of the strap evolves the surface moves towards being at the same stress level. Because of

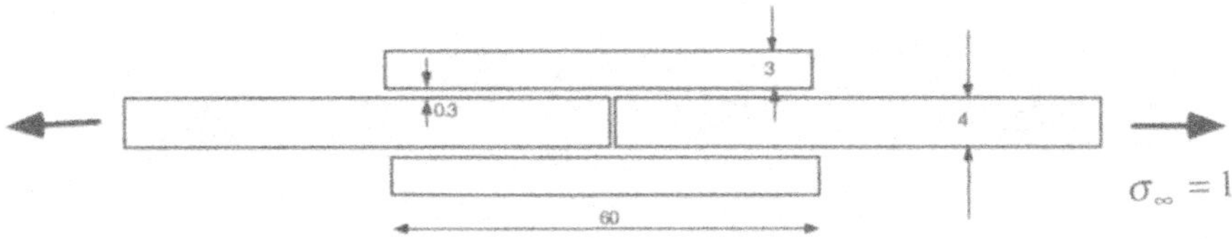

Fig. 8.13. Double strap butt joint.

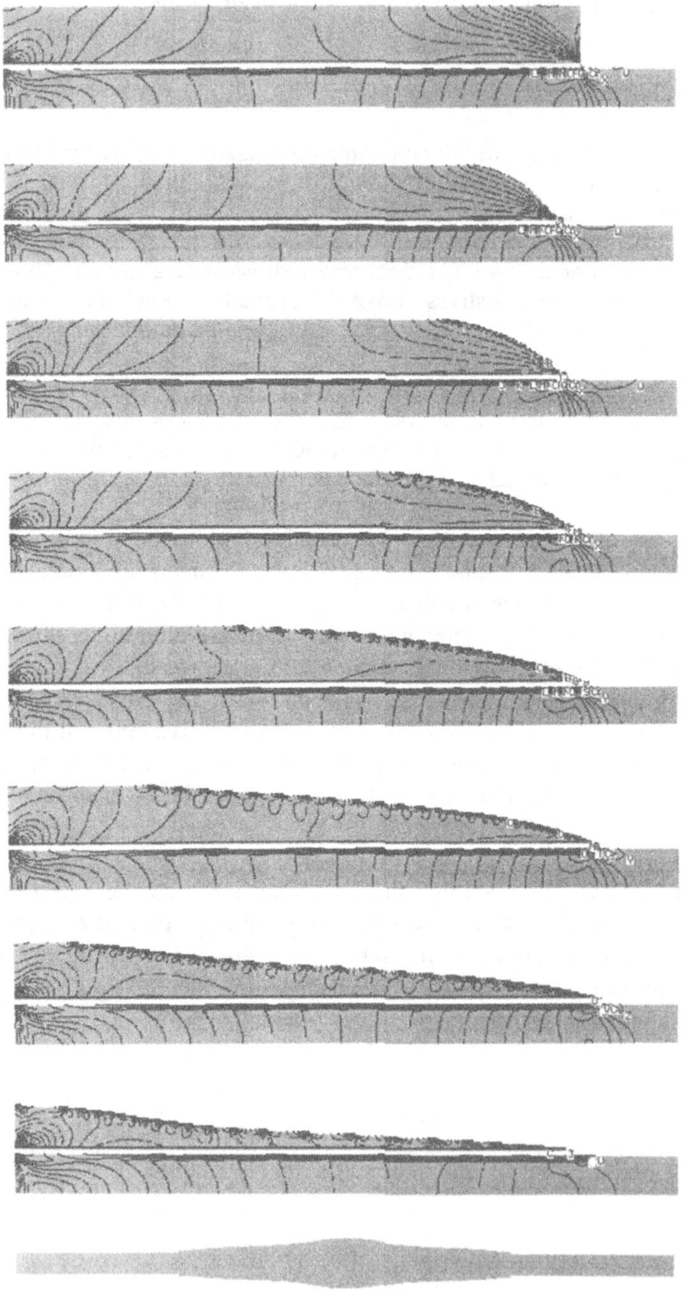

Fig. 8.14. Evolving shapes for the butt joint where only the strap can evolve.

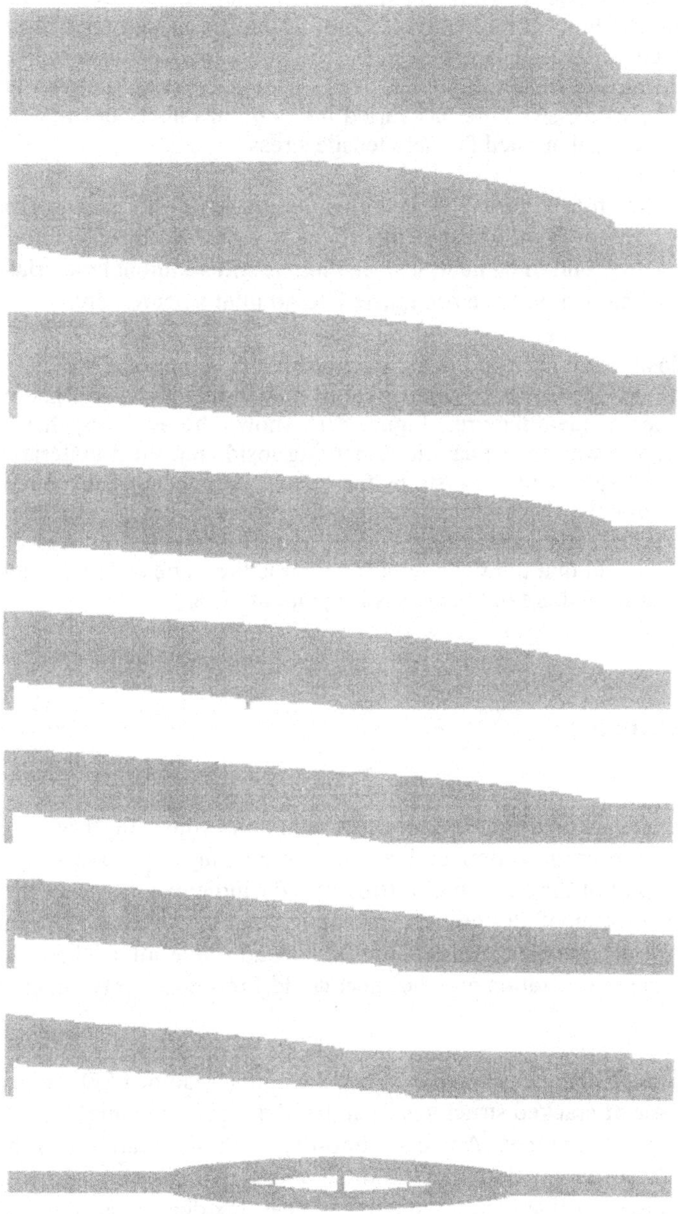

Fig. 8.15. Evolving shapes for the butt joint where both strap and sheet can evolve.

the jaggy nature of the surface with the ESO algorithm there is the clustering of contours at the steps. However the stress analyst can detect that the stress throughout the bulk of the strap material is almost constant apart from the bottom left hand corner which is not available for removal. It is also worth stating that even with the significant reduction in the volume of strap material the maximum stress in the whole structure has varied little and has never been greater than 11% above the nominal applied far field tensile stress.

The final sketch in Figure 8.14 is the whole joint with the best evolved shape for the straps. Although the stress in this shape is only 8% above the nominal one the joint can not be said to be neutral since there is still additional material needed over and above what would have been there had no joint been required.

As a curiosity extension to the above analysis where only the straps are allowed to evolve we let the sheet material evolve in a limited way, forgetting about the practicalities of manufacture. Figure 8.15 shows the evolving shape under these conditions and we can see cavities emerging inside the sheet material. Because the load is being forced into the straps there is a component of the force in the strap acting towards the sheet which produces the pillars across the gap that forms in the sheet material. It is interesting to note that this shape produces a joint that is almost neutral in that the volume of material is the same as there would have been had no joint been there but the stress is up about 10%.

8.4 Conclusion

This chapter has presented a sample of situations where the ESO process can be applied to the optimization problem of minimizing the maximum stress on the surface or part of the surface of a structure. By judicious design of the FEA model and the selection of design and non-design material the analyst can achieve a variety of outcomes such as the double butt adhesive joint where the straps had only a shape optimization and the sheet could form holes only emanating from the mid-plane.

Other research is under way to examine the application of ESO shape optimization to the repair of cracked structures such that the repair is neutral from both strength and mass points of view. Another interesting field of research is to simultaneously perform heat conduction and thermal stress evolutionary shape optimization whereby the shape is optimum in terms of heat flux density as well as stress.

It is interesting to note that the ESO process for minimizing the maximum surface stress is the computational analog of an experimental stress concentration minimization process using a photoelastic method (Heywood 1969; Durelli and Rajaia 1981). In this experimental method the initial design is studied in the photoelastic apparatus. High stress concentration is indicated by closely spaced fringes. A scraping device gently removes material from the low fringe density area of interest. Using the visual feedback the operator builds up sufficient skill to be able to remove surface material in such a way as to reduce the stress concentration.

8.5 References

Backlund, J. and Isby, R. (1988) Shape optimization of holes in composite shear panels. In: G.I.N. Rozvany and B.L. Karihaloo (eds.), *Structural Optimization*, Kluwer Academic Publishers, Dordrecht, 9-16.

Ding, Y. (1986) Shape optimization of structures: a literature survey. *Comput. Struct.* **24**(6), 985-1004.

Durelli, A.J. and Rajaia, K. (1981) Minimizing stress with photoelasticity. *Machine Design.* **53**(28), 125-130.

Engineering Science Data Unit (1969) *Elastic Stress Concentration Factors: Geometric Discontinuities in Flat Bars or Strips of Isotropic Materials*, Amended 1979, 251-259 Regent St., London, W1R 7AD, U.K.

Falzon, B.G. (1996) *An Investigation into the Buckling and Postbuckling Behaviour of Hat-Stiffened Composite Panels*, PhD Thesis, Department of Aeronautical Engineering, University of Sydney.

Falzon, B.G., Steven, G.P. and Xie, Y.M. (1996) Shape optimization of interior cutouts in composite panels. *Struct. Optim.* **11**, 43-49.

Haftka, R.T. and Grandhi, R.V. (1985) Structural shape optimization: - a survey. *AIAA-85-0772*, AIAA/ASME/ASCE/AHS 26th Structural Dynamics and Materials Conf., Orlando, Florida.

Heywood, R.B. (1969) *Photoelasticity for Designers*, Pergamon, Oxford.

Peterson R.E. (1974) *Stress Concentration Design Factors*, John Wiley, New York.

Tsai, S.W. (1988) *Composite Design*, 4th edn., Think Composites, Dayton.

Chapter 9 ESO Computer Program Evolve97

by O. M. Querin

This book has so far given you an insight into the ESO method and on how it is applied to carry out optimization of various structural problems. A difficulty that can arise from a book such as this is that, although the reader can understand how the process works, it is very hard to apply it on demand. This comes about because the reader needs to have access to various computer programs such as Finite Element Analysis software, optimization software, and a good programming background in order to implement what is explained. Fortunately this book includes with it two computer programs that will allow you to test and use the ESO method. The first is ESOFRAME as listed in the appendix of Chapter 7 and the second is Evolve97. Evolve97 is the first Evolutionary Structural Optimization program in the world which is developed in the environment of Windows 95 (Windows, Windows 95 and Windows NT are trademarks of Microsoft Corporation). Although relatively small in size (only 480kB), it will allow you to see how easy it is to use the ESO method to optimize structures. If you need more information about Evolve97 or wish to obtain future upgrades, you can have a look at our Web site: **www.ae.su.oz.au/wwwdocs/eso1.html** or simply send us an E-mail at: **ozz@aero.ae.su.oz.au** .

9.1 Introduction

The computer diskette that comes with this book includes the ESO software Evolve97. This is a cut down version of the full program, with restrictions on the maximum Finite Element mesh size that can be used. However, what is provided is still powerful enough to give you a hands-on approach to the ESO method without having to do any programming.

This chapter deals with the operation of Evolve97 and can be viewed as a User Guide for the program. As well, it deals with some examples and explains how to use Evolve97 to carry out the optimization of a structure. This version of the program is limited in the number of elements that can be used in a model. The maximum mesh size is 55×88, totalling 4840 plate elements. As well this program can only optimize for one load case and one criterion (the von Mises stress). The full version of the program has no restrictions and can be used for all element types (beam, plates and bricks), multiple load cases and multiple criteria, as well as other solution types such as natural frequency and buckling optimization.

9.2 System Requirements and Installation of Evolve97

Evolve97 is a 32-bit application for Windows, so you should have Windows 95 or Windows NT installed in your computer to run Evolve97. You should also have at least 16 MB of RAM installed in your machine and no less than 50MB of free Hard-Disk space. Evolve97 may not run if your system is smaller than this. In Windows 95's **'Display Properties' - Settings** section, the **'Color Palette'** must be set to **'High Color (16bit)'**. Evolve97 has been configured for this number of colours and may not work efficiently if lower colour settings are selected.

To install Evolve97, follow these steps:

1 Start your computer, with either Windows 95 or Windows NT.
2 Insert the disk accompanying the book into the floppy disk drive.
3 Using the Windows Explorer or File Manager, create a folder in your computer hard disk and name it 'Evolve97Program'.
4 Copy the program Evolve97, shown in Figure 9.1, to the folder you just created in your computer called 'Evolve97Program'.
5 Create a Shortcut to the Evolve97 program, copy it using the Edit command, then navigate to the 'Programs' folder located in: '\Windows\Start Menu\Programs', and Paste the Shortcut.
6 Installation is now complete. Remove the supplied floppy from your computer.
7 To start Evolve97, select Start from the Windows Taskbar, select Programs, and select Shortcut to Evolve97.exe. Alternatively from within Windows Explorer or File Manager, double click on the icon shown in Figure 9.1.

Fig. 9.1. Evolve97 icon.

9.3 Overview of the Evolve97 Interface

When Evolve97 is started, the interface of Figure 9.2 appears. This is similar to most Windows applications as it has most of the usual features. The interface is divided into five sections: (A) Title bar, (B) Menu bar, (C) Button bar, (D) Display area and (E) Status bar.

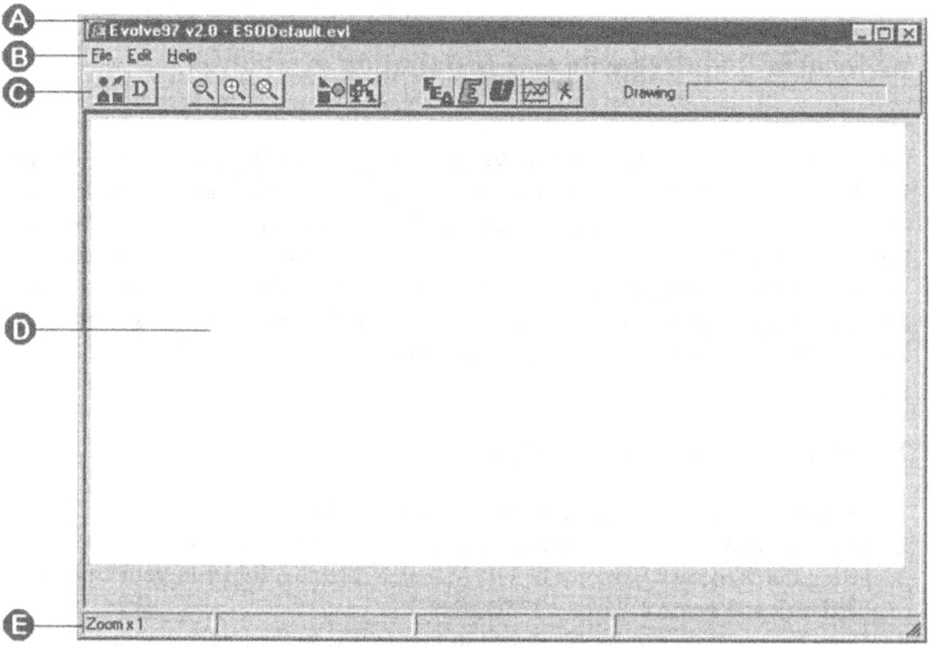

Fig. 9.2. Evolve97 Interface.

9.3.1 Title Bar

The Title Bar (A) displays the name of the program 'Evolve97' together with the name of the current structure file. On the right hand side it also has the standard Windows 95 buttons to Minimize, Maximize and Close the program.

9.3.2 Menu Bar

The Menu Bar (B) has three menus that you can select from, File, Edit and Help.

If File is selected, the drop-down list of Figure 9.3 is displayed. These components have their usual meaning as in other Windows applications.

New
Open
Save as

Print
Print Setup

Exit

Fig. 9.3. File drop down menu list.

- New, allows you to type the file name of a new structure to be optimized.

- Open, as the name suggests, opens an existing structure file that has been optimized or that you are still working on.

- Save as, allows you to save the structure you are working on under a different name.

Note : *A Save menu item has not been provided. This is because in every step of a structure's development and evolution, Evolve97 automatically saves ALL corrections, modifications, loads, freedom conditions, properties and all other parameters in files with the name specified in New but with different extensions. The names of these extensions will be given in Section 9.5.*

- Print, selecting this will print the viewing screen directly to the printer. In this version of Evolve97, you are unable to print to a file, you must print directly to the printer. To print to a file, you can save the Display Area using the 'Save Display Area to Bitmap File' button (which will be explained later), and then use a Windows application to print to a file.

- Print Setup, allows you to set up and configure the printer connected to your computer.

- Exit, will exit the program.

If Edit is selected from the menu bar, the drop-down menu list of Figure 9.4 is shown:

Cut	Ctrl + X
Copy	Ctrl + C
Paste	Ctrl + V

Fig. 9.4. Edit drop down menu list.

- Cut, allows you to cut the entire viewing frame to the clipboard as a bitmap image.
- Copy allows you to copy the entire viewing frame to the clipboard as a bitmap image.
- Paste, allows you to paste an image from the clipboard into the Display Area.

Note: *If you draw the image of a structure in another application such as Paint, the design domain must be drawn using the colour combination RGB = 0,127,0 (Green) and the non-design domain must be drawn using the colour combination RGB = 255,127,0 (Orange).*

Selecting Help from the menu bar shows the drop-down menu list of Figure 9.5:

Help
About

Fig. 9.5. Help drop down menu list.

- Help, gives you access to the On-Line help. However this feature is not present in this version of Evolve97.
- About, gives you some information about Evolve97 together with E-mail and Web site addresses to obtain future upgrades.

9.3.3 Button Bar

The Button Bar (C) contains 12 buttons and a progress bar. This section will explain what each button does .

This is the **'Select items to draw'** button. By pressing this, it opens the dialogue box of Figure 9.6. Clicking on the check boxes, you can select those items of the structural model you wish to draw. You can select from Nodes, Freedoms, Point Loads and Plates. You can draw any of them individually by having only that box checked or you can select to have all items drawn as is shown in Figure 9.6. There are also two radio buttons present. These allow you to select drawing the plate properties or the plate stresses. The plate stress radio button is however only enabled after a Finite Element Analysis has been carried out.

Note: *By using the Tab key from your keyboard, you can toggle between items in ALL of the dialogue boxes created by Evolve97.*

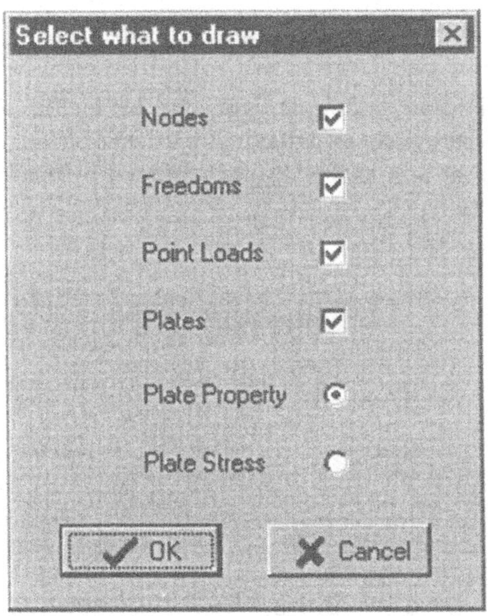

Fig. 9.6. Dialogue box for selecting items to draw.

This is the **'Draw'** button. After you have selected what to draw, by pressing this button you can draw the structure in the Display Area (D) (see Figure 9.2).

This is the **'Zoom Out'** button. By pressing this button, it reduces the size of the Display Area (D) by a factor of 0.1, every time this button is pressed. The minimum Display Area size is 0.2 times the original size.

This is the **'Zoom In'** button. By pressing this button it increases the size of the Display Area (D) by a factor of 0.1, every time this button is pressed. The maximum Display Area size is 2 times the original size.

This is the **'Zoom Factor'** button. By pressing this button, it opens the dialogue box of Figure 9.7, where you can type the Zoom factor in the range from 0.2 up to 2. This will change the size of the Display Area (D) by the factor you typed after you press the OK button.

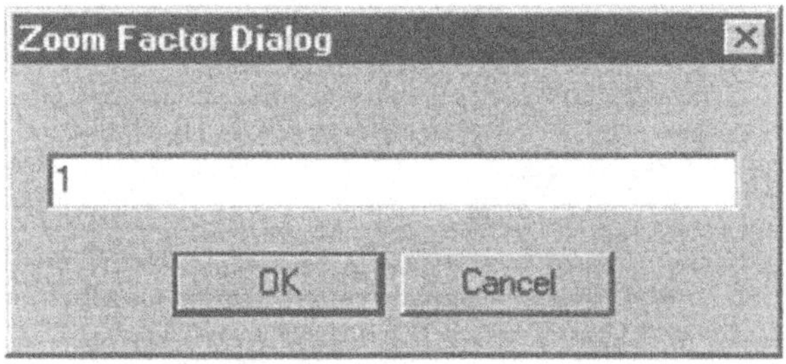

Fig. 9.7. Dialogue box for zoom factor.

This is the **'Drawing Tools Menu'** button. By pressing this button, it opens the 'Drawing tools button bar' of Figure 9.8. Section 9.3.6 will explain the function of all the buttons in this bar.

Fig. 9.8. Drawing tools menu bar.

This is the **'Mesh Structure and ESO Parameter Menu'** button. By pressing this button, it opens the 'Mesh structure and ESO parameter button bar' of Figure 9.9. Section 9.3.7 will explain the function of all the buttons in this bar.

Fig. 9.9. Mesh structure and ESO parameter menu bar.

This is the **'Carry out Finite Element Analysis'** button. Pressing it, opens the button bar of Figure 9.10. Section 9.3.8 will explain the function of the two buttons and of the progress bars.

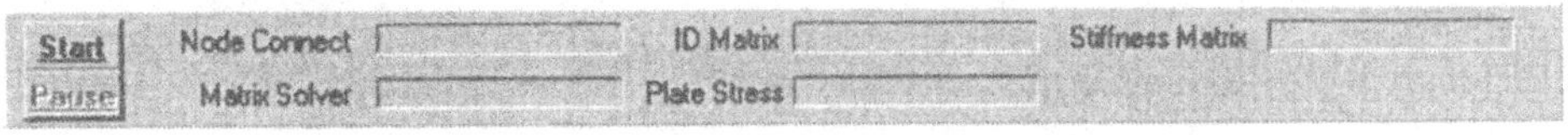

Fig. 9.10. Finite Element Analysis button bar.

This is the **'Evolve structure'** button. By pressing this button it opens the button bar of Figure 9.11. Section 9.3.9 will explain the function of the two buttons and of the progress bars.

Fig. 9.11. Evolve structure button bar.

This is the **'UnEvolve structure'** button. By pressing this button it opens the dialogue box of Figure 9.12 which allows you to UnEvolve the structure to a specific Rejection Ration (*RR*) value.

To UnEvolve a structure simply type the specific *RR* value in the edit box of Figure 9.12. Once you have viewed the unevolved structure and either re-calculated its stresses by carrying out FEA or saved the structure's image using the 'Save Display area to bitmap' button (see below), you should always **Restore** the structure to the last Rejection Ratio reached by the Evolution process. To do this press the Restore button shown in Figure 9.12, and then press the OK button.

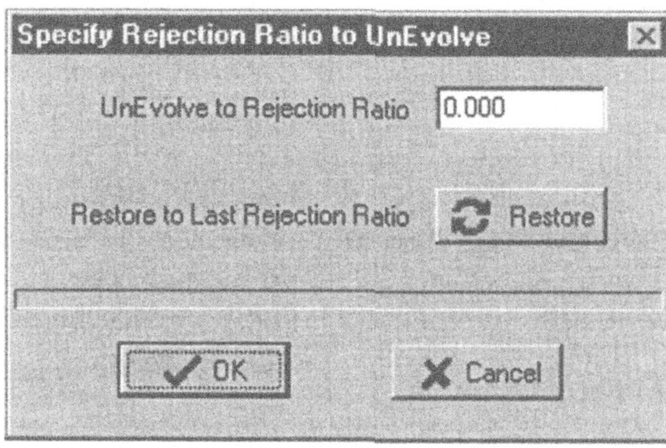

Fig. 9.12. Dialogue box for specifying rejection ratio to UnEvolve structure.

This is the **'Plot Evolution History of Structure'** button. By pressing this button it opens the dialogue box of Figure 9.13 which allows you to specify the parameters and axis limits of the plots that will display relevant parameter changes during the evolution history of the structure.

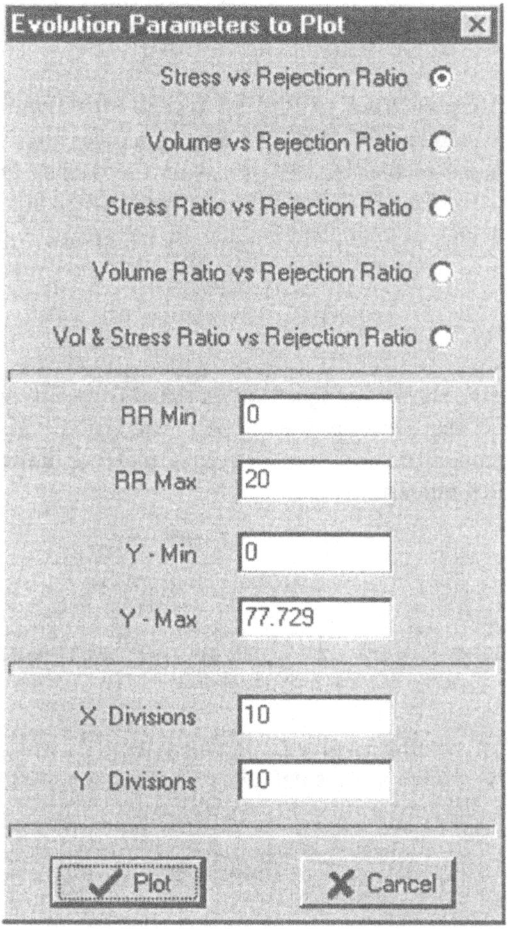

Fig. 9.13. Evolution parameters to plot dialogue box.

As is clearly denoted in the labels of the dialogue box of Figure 9.13, you have a choice of plotting graphs of Stress vs *RR*, Volume vs *RR*, Stress ratios vs *RR*, Volume ratios vs *RR* and all ratios vs *RR*. All you need to do is to select the appropriate radio button.

The program will then automatically select for you the axis range limits. If you want to change any of them simply type the appropriate value in the relevant edit box. When you are satisfied with the limit range, just press the Plot button and the selected graph will be plotted in the Display area. If you want to save this graph use the 'Save Display Area to Bitmap Image File' button (see below).

This is the **'Save Display Area to Bitmap Image File'** button. By pressing this button, it opens the standard Windows 95 Save dialogue box, where all you need to do is type the name of the file that you want the display area to be saved to. Since the saved file format is Windows bitmap, you will be able to open the image in any image editor such as Paint, and include the image in any document.

9.3.4 Display Area

The Display Area (D), shown in Figure 9.2, is the only area of the screen where the structure is drawn. It also serves as the region of the visible screen where you draw the original maximum structural domain that is to be optimized. And this is the area that is saved as a bitmap file.

9.3.5 Status Bar

The Status Bar, shown in Figure 9.14, is where different types of information are displayed.

When you first start running Evolve97, the only item shown in the Status Bar, is the Zoom factor (A), which always defaults to 1 when a new structure is generated or opened.

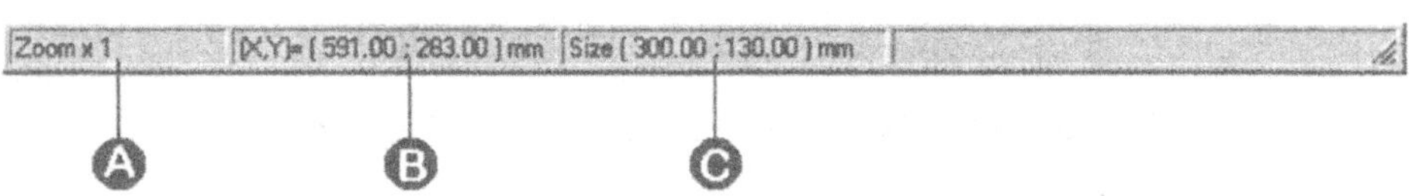

Fig. 9.14. Status bar showing mouse coordinate and size of object drawn.

When you enter the Drawing Tools menu (see Section 9.3.6), or the Mesh structure and Parameter set up menu (see Section 9.3.7), the second cell (B) of the Status bar indicates the horizontal and vertical mouse coordinates.

When an object is drawn on the Display Area, the third cell of the Status bar (C) shows the horizontal and vertical dimensions of the object. This is only done for the Rectangle and Ellipse. For this version of Evolve97 the Polygon shape is not specified in this cell of the status bar.

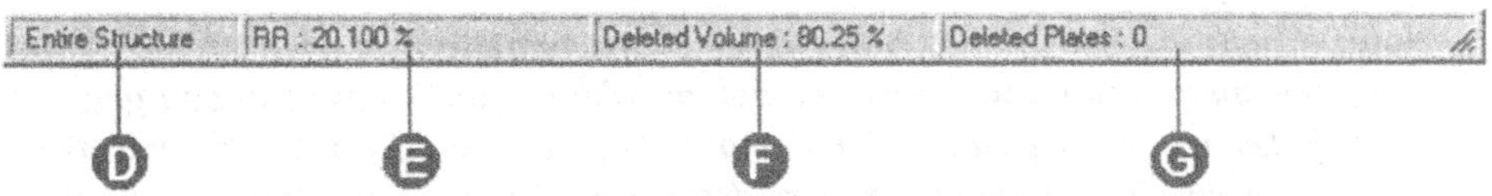

Fig. 9.15. Status bar showing the region of the structure that can be evolved, current *RR*, percentage of deleted volume and the number of plated deleted in the last iteration.

When the structure is evolving, the status bar in the first cell (D) shows whether the evolution process is taking place on the entire structure or only on the perimeter. The latter is termed Nibbling and is analogous to shape optimization, which has been explained in Chapters 2 and 8.

The second cell (E) shows the current Rejection Ratio (*RR*) of the evolution process. The third cell (F) shows the percentage of the original volume of the structure that has been removed. The fourth cell (G) indicates the number of plate elements that were removed in the last iteration.

When a structure is unevolved, the fourth cell shows the Rejection Ratio which the structure has been unevolved to.

9.3.6 Drawing Tools Button Bar

In Section 9.3.2 it has been shown that by pressing the 'Drawing Tools Menu' button, you are able to access the 'Drawing tools button bar' of Figure 9.8. This section explains what each of the nine buttons of that button bar does.

This is the **'Use Design Property'** button. When this button is in the pressed down mode, all shapes drawn in the Display Area will be green in colour. The areas drawn in green are the only ones that can be removed in the evolution process.

This is the **'Use NON-Design Property'** button. When this button is in the pressed down mode, all shapes drawn in the Display Area will be orange in colour. The areas drawn in orange **cannot** be removed in the evolution process.

Note: *If both the Design and Non-Design property buttons are not pressed down, the shape drawn will have the colour white. This is how you can generate holes in your original design domain. Alternatively, if you made a mistake in what you are drawing, you can draw over the incorrect region with any of the shape tools with both Design and Non-Design property buttons pressed down to delete that region.*

This is the **'Draw Rectangle'** button. When this button is in the pressed down mode, you can draw rectangular shapes in the Display. If you hold down the Shift key as you draw a rectangular shape, a square is drawn.

This is the **'Draw Ellipse'** button. When this button is in the pressed down mode, you can draw elliptical shapes in the Display Area. If you hold down the Shift key as you draw an elliptical shape, a circle is drawn.

This is the **'Draw Polygon'** button. When this button is in the pressed down mode, you can draw a polygon in the Display Area.. The polygon can have up to a maximum of 250 vertices. To draw a polygon you start by clicking inside the Display Area where you want the first vertex to appear, then continue to move the mouse and click the left mouse button in all the locations where you want vertices of the polygon to appear. As you do this, lines are drawn between the vertices consecutively. To complete the polygon, double click the left mouse button on the last vertex point and the polygon shape will be closed and filled with material of selected property.

This is the **'Clear Display Area'** button. When this button is pressed, the entire Display Area is cleared. There is no undo button, so be very careful not to press this button by mistake. An alternative way of clearing the display area, is to use the

Cut command in the Edit menu. This way if you find you have made a mistake in clearing this area, you can always bring it back by using the Paste command.

This is the **'Specify Length of Display Area'** button. When this button is pressed, the dialogue box of Figure 9.16 appears and you can then type in the edit box the width of the display area in millimetres.

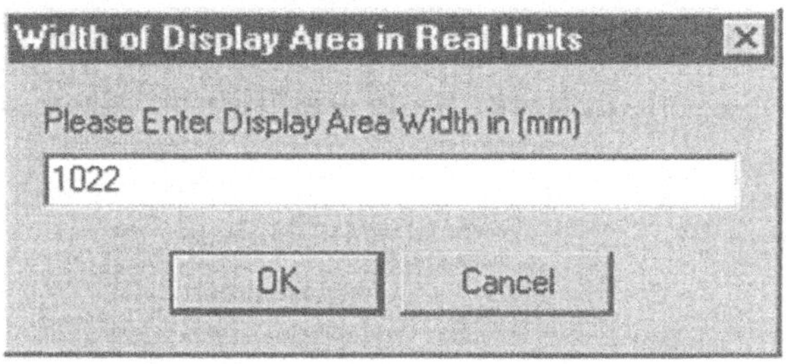

Fig. 9.16. Dialogue box where the width of the Display Area is specified.

This is the **'Define Finite Element Mesh Density'** button. When this button is pressed the dialogue box of Figure 9.17appears. This dialogue box allow you to select from five pre-defined mesh densities that Evolve97 will discretize the display area into.

Selecting an appropriate mesh for a structure is a delicate matter since, although a very fine mesh will give you very accurate, the time to complete the optimization process may be very long. On the other hand, a model with a coarse mesh will be optimized very quickly but the final shape may be very rough. As you become more familiar with Evolve97 and ESO, you will gain experience on which mesh density to use. To begin, however, we suggest you use the coarsest mesh density. This will give you results very quickly so that you can witness how the ESO method works without waiting too long.

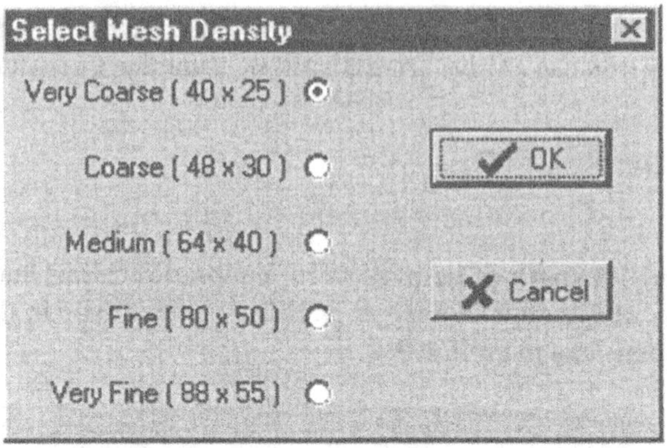

Fig. 9.17. Dialogue box where the density of the Finite Element Mesh is selected.

This is the **'Create Finite Element Mesh'** button. By pressing this button, the dialogue box of Figure 9.18 appears. When you select Start, Evolve97 will create a Finite Element mesh of the object drawn in the Display Area with a mesh density specified by the above 'Define Finite Element Mesh Density' button.

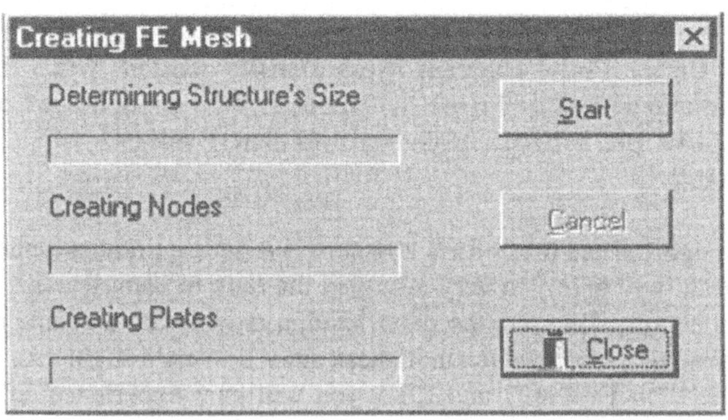

Fig. 9.18. Dialogue box used to create the Finite Element Mesh of the object drawn in the Display Area.

9.3.7 Mesh and ESO Parameter Button Bar

In Section 9.3.2 it was shown that by pressing the 'Mesh Structure and ESO Parameter Menu' button, you are able to access the 'Mesh structure and ESO parameter' button bars of Figure 9.9. This section explains what each of the ten buttons does.

This is the **'Design Property'** button. By pressing it, the dialogue box of Figure 9.19 appears. You must use the units specified in the dialogue box for the material property of the design domain. Figure 9.19 shows typical values for steel.

Note: *Evolve97 sets the Design and Non-Design material properties of all new structures to the default values of Steel shown in Figure 9.19.*

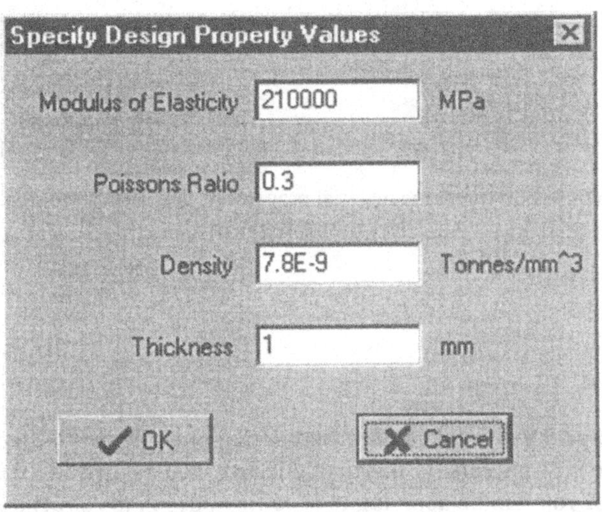

Fig. 9.19. Dialogue box where you can specify the material property of the structure that can be evolved.

This is the **'Non-Design Property'** button. By pressing it, the dialogue box of Figure 9.20 appears. Similar to the Design Property menu, it is used to specify the material property of the non-design region of the structure.

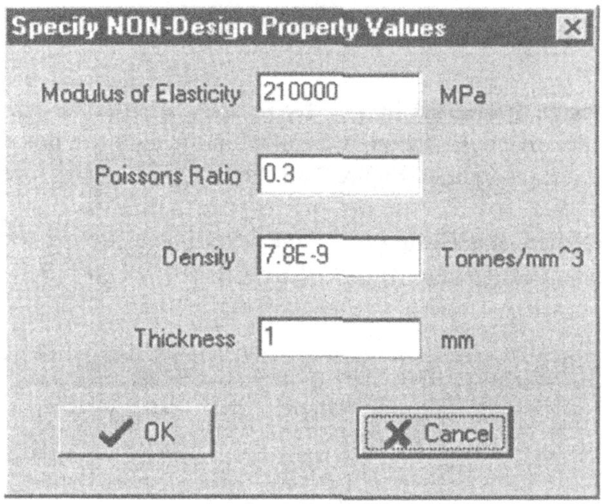

Fig. 9.20. Dialogue box where the material property of the parts of the structure that cannot be evolved is specified.

This is the **'Specify Applied Load'** button. By pressing this button, the dialogue box of Figure 9.21 appears. In this dialogue box, you can specify the magnitude of the forces that you will be applying to specific nodes in the structure. As well you can also specify accelerations to the entire structure. Be aware that while forces are applied to individual nodes, the accelerations are applied to every plate element of the structure.

Note : *The unit for an applied Force must be Newtons (N) and the units for Acceleration mm/s^2. If you the earth's gravitational acceleration is to be applied to a structure, the magnitude you should type is -9800 mm/s^2 in the edit box of Acceleration-Y.*

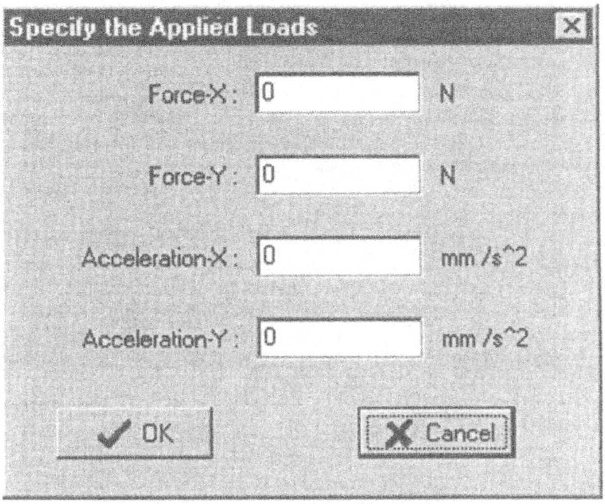

Fig. 9.21 Dialogue box that allows you to type the loads applied to the structure.

This is the **'Apply Loads to Node'** button. When this button is in the pressed down position, by moving the mouse to a node and clicking the left mouse button, the forces specified above will be added to that node.

If you make a mistake in applying a load, you can delete it by setting both Force-X and Force-Y to zero in the 'Specify Applied Load' dialogue box and then applying the setting to that node.

This is the **'Specify Node Freedom'** button. By pressing it, the dialogue box of Figure 9.22 appears. This allows you to specify freedom conditions to all the nodes.

Fig. 9.22. Dialogue box where you select the global and individual node freedom conditions.

Global freedoms are applied to all nodes that you have not specified freedoms to. For almost all structures that you optimize you should leave the global freedoms set to 'All Displacements Free', which is the default condition and can be seen in Figure 9.22.

For the Individual node settings you can choose from four different freedom conditions. A node can have 'X-Displacement Only', 'Y-Displacement Only', 'All Displacement Fixed', or 'All Displacements Free'.

This is the **'Apply Freedom Condition to Node'** button. When this button is in the pressed down position, moving the mouse to a node and clicking the left mouse button, you can apply the freedom conditions specified above to that node.

If you make a mistake in applying the freedom condition, you can set the Individual Node Settings to All Displacements Free and reapply the freedom condition to that node.

This is the **'Delete Individual Plate Elements from Mesh'** button. When this button is in the pressed down position, moving the mouse to a plate and clicking the left mouse button, you can delete that plate from the structure.

This is the **'Delete Entire Finite Element Mesh'** button. When this button is pressed the dialogue box of Figure 9.23 appears and asks you if you really want to delete the entire structure. If you press the 'Yes' button the structure and all of its associated files will be deleted. If that is not what you intended to do, press 'No' or 'Cancel' button.

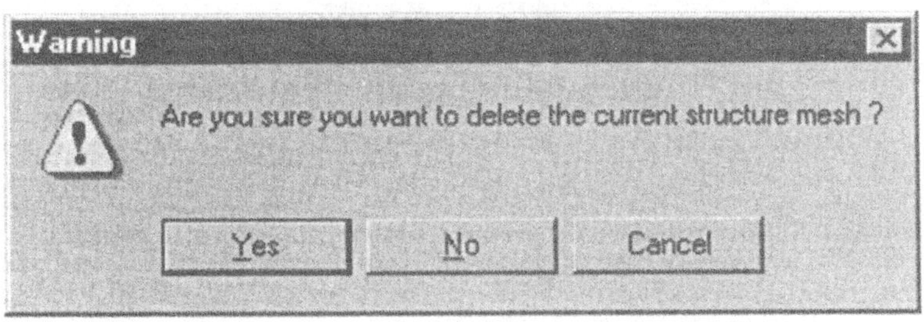

Fig. 9.23. Warning dialogue box that asks you if you want to delete the current structure mesh.

This is the **'Draw New Plates to Mesh'** button. When this button is pressed the dialogue box of Figure 9.24 appears, giving you a choice in the property type of the plate element that you will be adding to the structure. You have a choice from Design Property (Green) or Non-Design property (Orange). To add a plate with the chosen property, simply click the left mouse button at the centre of four nodes where you want a plate to be added.

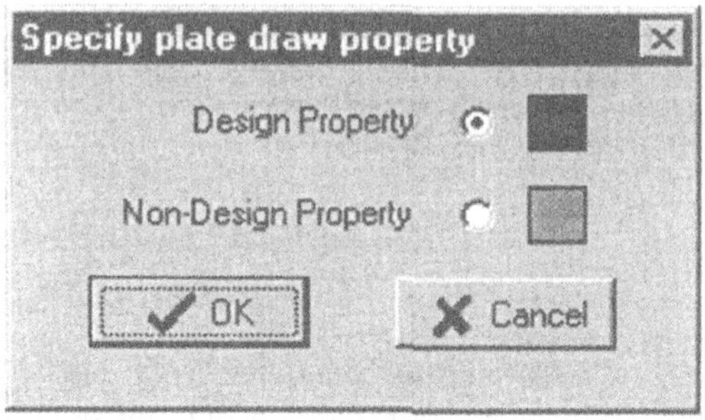

Fig. 9.24. Dialogue box for specifying the property of plates to be drawn.

This is the **'Specify ESO Parameters'** button. When this button is pressed, the dialogue box of Figure 9.25 appears. The values shown in this figure are the default settings specified by Evolve97 for all structures.

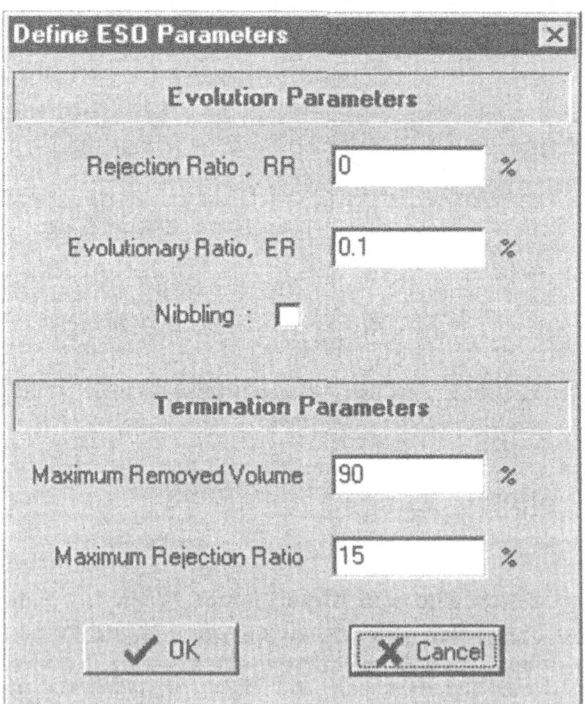

Fig. 9.25. Dialogue box for defining ESO parameters.

For the majority of structures that you will optimize, you do not need to change these parameters.

After the evolution process is started, the *RR* edit box will show the current Rejection Ratio. The reason for doing this is that you may wish to unevolve the structure to a specific *RR* value and continue from there with a different *ER* value.

Although you can select any values for the Rejection Ration and the Evolution Ratio, values greater than 1% for both will not always give satisfactory results. The recommended default values of *RR* = 0% and *ER* = 0.1% are sufficiently small and will give very good results.

The Nibbling check box allows you to force the ESO process into carrying out shape optimization. This is done by removing elements from the structural boundaries only, with no internal holes being created.

In order for the ESO process to terminate before all the material is removed, two termination parameters have been provided: Maximum Removed Volume and Maximum Rejection Ratio. Figure 9.25 shows the default values.

Depending on the mesh density used, you may have to increase the Maximum Rejection Ration value to 20 or 25%. However you do not need to do this before the process is started. If it reaches the 15% limit and the process is stopped, you can enter this dialogue box again, increase this value and re-start the evolution process again. It will start from where it stopped.

9.3.8 Finite Element Button Bar

In Section 9.3.2 it has been shown that by pressing 'Carry out Finite Element Analysis' button, you are able to access the 'Finite Element Analysis' button bar of Figure 9.26. This section explains what you need to do to carry out a finite element analysis of the structure.

When you enter this button bar you will see Start and Pause buttons as shown in Figure 9.26. The Start button is enabled, and is the only one you can press at the beginning. Once the finite element analysis is started, the Start button becomes disabled and the only button that you can press is the Pause button, which can be used to terminate an analysis it you want to. The Progress Bars show you at what stage of the analysis Evolve97 is while it is running.

When all five progress bars are full and the Start button becomes enabled again, you can view the von Mises stress distribution by pressing the 'Select Element to Draw' button and selecting Plate stresses, and then pressing the Draw button.

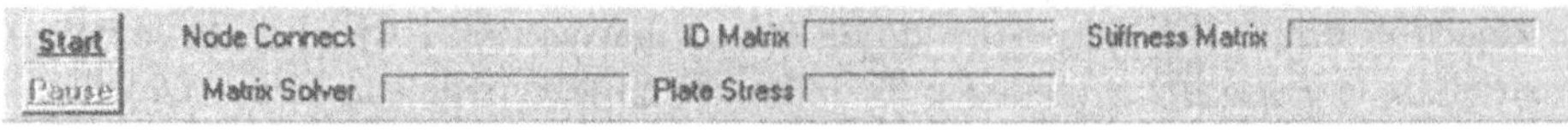

Fig. 9. 26 Finite Element Analysis button bar.

9.3.9 Evolve Structure Button Bar

In Section 9.3.2 it has been shown that by pressing 'Evolve structure' button, you are able to access the 'Evolve structure' button bar of Figure 9.27. This section explains what you need to do to evolve a structure.

In a similar way to the 'Finite Element Analysis' button bar, you have two buttons: Start and Pause. The Start button is used to start the optimization process. Once this happens the Start button is disabled and the Pause button becomes active.

The two progress bars provide you with information on what step of an evolution cycle Evolve97 is at, either FEA or ESO.

By pressing the Pause button you can terminate the evolution process at any stage. You can also re-start the evolution process by pressing Start again.

Fig. 9. 27 Evolve structure button bar.

Every time an iteration cycle is completed, Evolve97 will draw for you in the display area the current structural shape. If you want the drawing process to be expedient, just select Plate Elements and Plate Stress, from the Select Element to Draw button. And you will be quite mesmerized by how the stresses in the structure change, and before long you will be trying to outguess Evolve97 as to which element to remove. One point to note it that the stress field that you see, is the one for the structure at the end of previous iteration. The stress field for the current structure is yet to be determined by the next FEA.

9.4 To Get Started

This section has been written to guide you through the steps necessary to construct a model and then evolve it. The steps are:

1. Drawing a structural domain.
2. Generating the finite element mesh.
3. Specifying the nodal forces, freedoms and material properties.
4. Specifying the ESO parameters.
5. Carrying out an FEA of the initial structure.
6. Evolving the structure.
7. Viewing the Evolution history.

9.4.1 Drawing a Structural Domain

To show how Evolve97 works, we will optimize the Michell type structure of Section 2.4.

- Start Evolve97
- Select File, and choose New.
- When the New dialogue box of Figure 9.28 appears type 'Michell Structure' and press OK.
- The name you just typed will now be shown in the Title Bar.
- Press the Drawing Tools Menu button.
- From the Drawing Tool button bar, press the Specify Length of Display Area button. In the dialogue box that appears, type the length 7000 and press the OK button.
- The Draw Rectangle and Use Design Property buttons should be in the down position (this is the default case every time you enter this menu). If either button is not down, press it.
- Move the mouse into the Display Area as close as possible to the coordinates (100, 3500).
- At this location, press the left mouse button, and while keeping it pressed move the mouse to the right and down until you create a rectangle of size 5000 x 2500 mm, try and get as close as possible to this, the shape you should get can be seen in Figure 9.29.

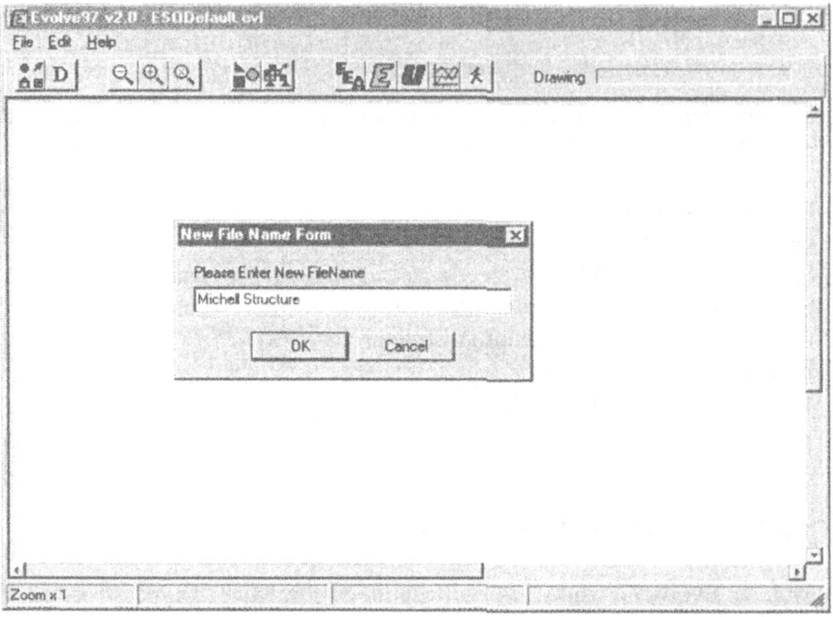

Fig. 9.28. Evolve97 interface showing new file name dialogue box.

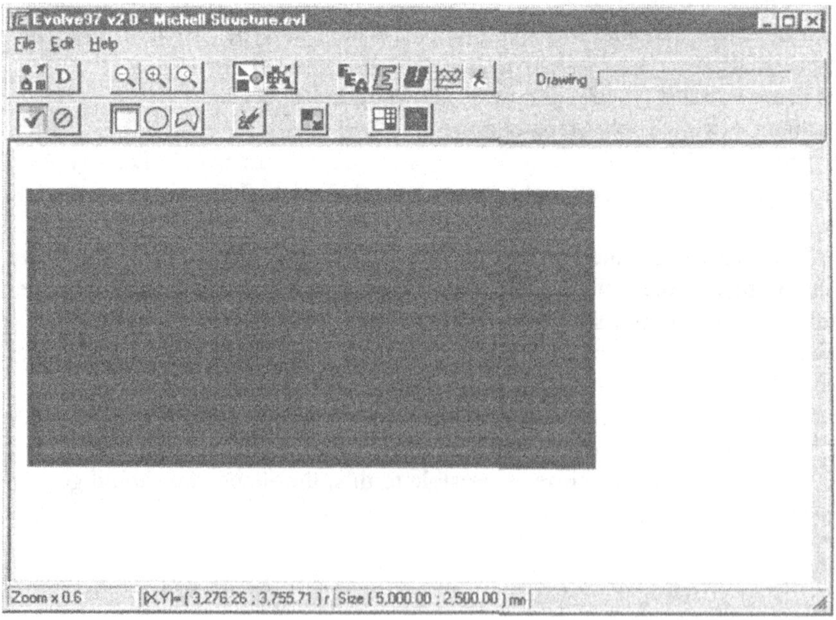

Fig. 9.29. Evolve97 interface showing drawn structural domain.

9.4.2 Generate Finite Element Mesh

Now that you have a drawing representing the design domain, follow these steps to generate the Finite Element mesh:

- From within the Drawing Tools Menu button, press the Define Finite Element Mesh Density button.
- The dialogue box of Figure 9.17 will appear, select the Medium (64×40) mesh, then press OK.
- Now press the Create Finite Element Mesh button.
- The Create FE mesh dialogue box of Figure 9.18 will appear, press Start.
- When the process is completed, press Close. The mesh of Figure 9.30 should be drawn. You now have a FE mesh of the design domain. You can now apply the freedom conditions , loads and material properties.

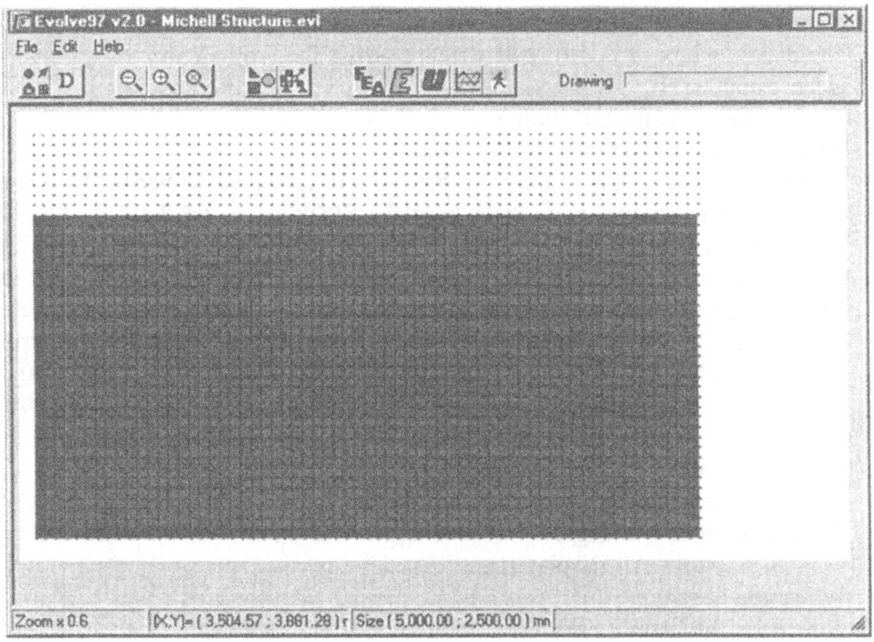

Fig. 9.30. Evolve97 interface showing the generated finite element mesh.

9.4.3 Specifying Properties, Loads, Freedoms and ESO Parameters

In order to do any work on the Finite Element mesh, you have to enter the Mesh Structure and ESO Parameter button bar. The buttons in this bar allow you to manipulate the mesh. To apply these parameters to the structure, follow these steps:

- Enter the Mesh Structure and ESO Parameter button bar by pressing the Mesh Structure and ESO Parameter Menu button.
- Press the 'Set Design Property' button.
- The dialogue box of Figure 9.19 should appear. Type the following values in the dialogue box:

Young's Modulus	:	100000 (MPa)
Poisson's Ratio	:	0.3
Density	:	0.0
Thickness	:	100 (mm)

- When finished, press the OK button.
- Press the 'Specify Applied Load' button.
- The dialogue box of Figure 9.21 will appear. In the Force-Y edit box, type the value -200000 (N).
- When done, press the OK button.
- Press the 'Apply Loads to Node' button.
- Move the mouse to the node in the bottom left hand side corner of the screen and read its horizontal coordinate from the Status bar. Add 2500 to this and move the mouse to the node in the bottom row of nodes closest to the coordinate you have calculated. This should be the central node along the bottom edge of the structure. Click the left mouse button, and a downwards pointing arrow should have been drawn. (You probably cannot see the arrow head but don't worry about that, the load has been recorded at that location). If you want to make absolutely certain that you have placed the load in the correct node. Using the Zoom Factor button, set a Factor of 2 and check that you have the correct node.
- Now press the 'Specify Node Freedom' button.
- The dialogue box of Figure 9.21 will appear. In the Individual Node Settings section, select the All Displacements Fixed button.
- When you have done this, press OK.
- Now press the 'Apply Freedom Condition to Node' button.
- Move the mouse to the node at the bottom left hand side corner of the structure and click the left mouse button.
- Then, move the mouse to the node at the bottom right hand corner of the structure and click the left mouse button.
- You should now see in front of you the image of the structure of Figure 9.31.
- Press the 'Specify ESO Parameters' button.

- The dialogue box of Figure 9.25 will appear with those same default values. For this example, these values need not to be changed. Just press the OK button in order for these values to be recorded for this structure.

9.4.4 Carry out FEA of Initial Structure

Although you may think it is not necessary, in order to verify that you have correctly specified all the parameters of the structure, it is a good practice to always carry out a FEA of the initial structure. There is nothing more frustrating than after hours of computation to obtain a wrong shape because of some incorrect settings.

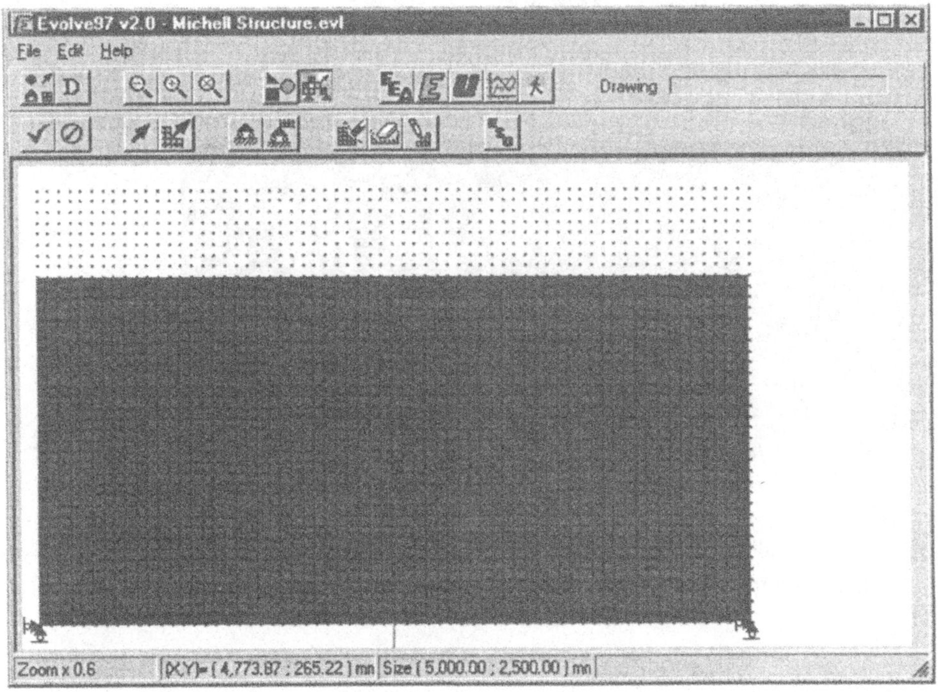

Fig. 9.31. Evolve97 interface showing the final structural domain.

To carry out a FEA of the structure, follow these steps:

- Press the 'Carry out Finite Element Analysis' button.
- Press the Start button.

- When the finite element analysis is finished press the 'Select element to draw' button.
- From the dialogue box that appears, select only the Plates from the Check buttons, and from the radio buttons select the Plate Stress, then press the OK button.
- Now press the Draw button.
- After a few seconds (this will be dependent on the speed of your computer) you should see the stress distribution in this structure of Figure 9.32.
- If you have obtained the same stress distribution, you are now ready to evolve the structure.

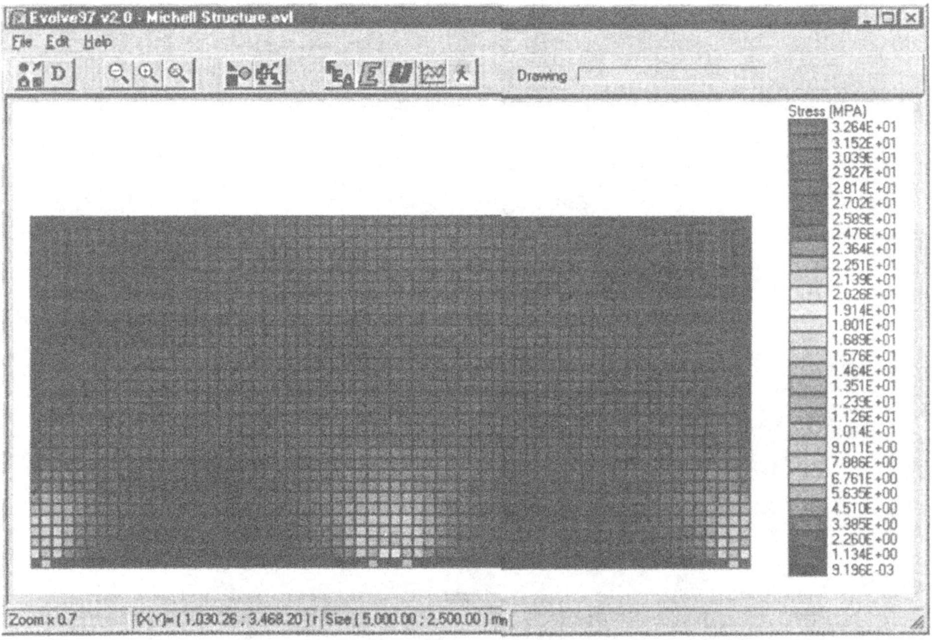

Fig. 9.32. Evolve97 interface showing the stress distribution in the initial structure

9.4.5 Evolving the Structure

In order to evolve the structure follow these steps:

- Press the Zoom Factor button, and set its value to 0.7.
- Press the 'Evolve structure' button.
- The Evolve structure button bar will appear, press the Start button.
- The structure should now start evolving right before your eyes.

- After one iteration, 24 plates, which is equal to 1.17% of the initial volume, should have been removed, giving you a structure that looks like the one in Figure 9.33.

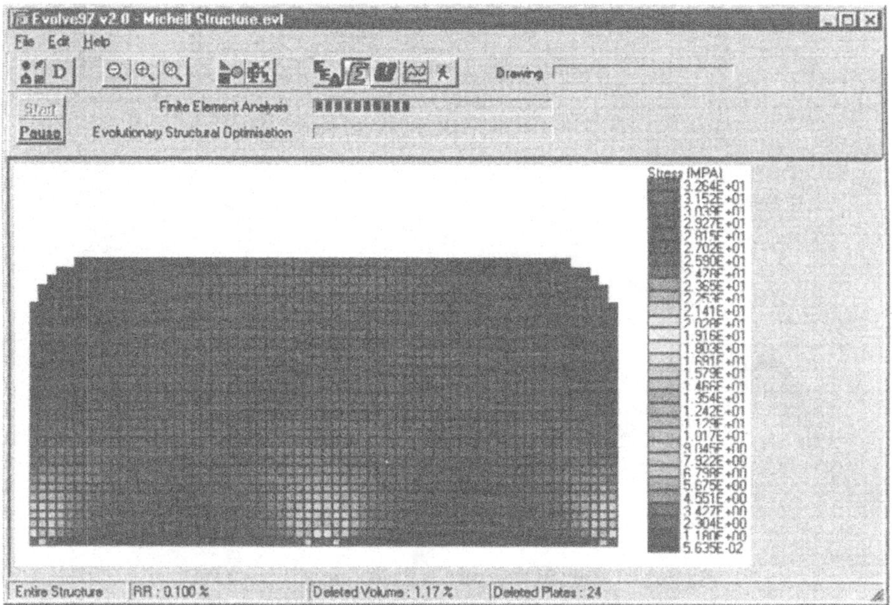

Fig. 9.33. Evolve97 interface showing the structure after one iteration.

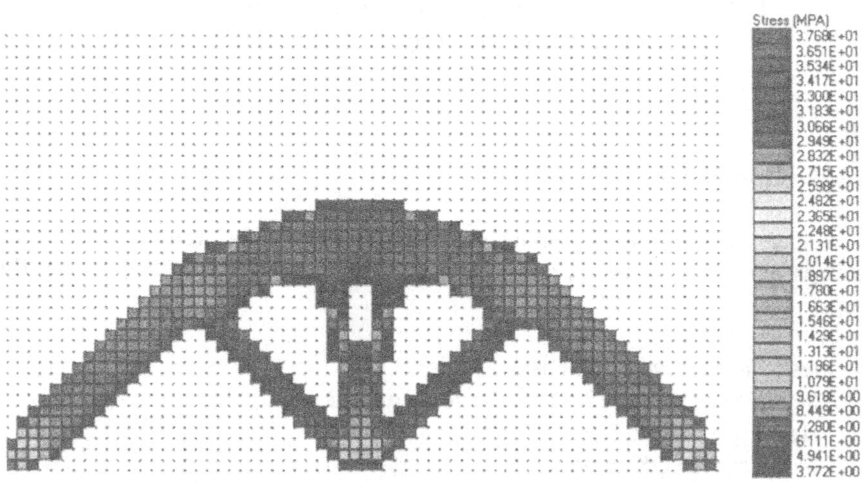

Fig. 9.34. Michell structure at Rejection Ratio equal to 10%.

- If you continue to watch the evolving structure, when the Rejection Ration reaches 10%, the structure should look like the one in Figure 9.34
- Continuing until the evolution is terminated, which is at Rejection Rate equal to 15%. The final structure should look like the one in Figure 9.35.

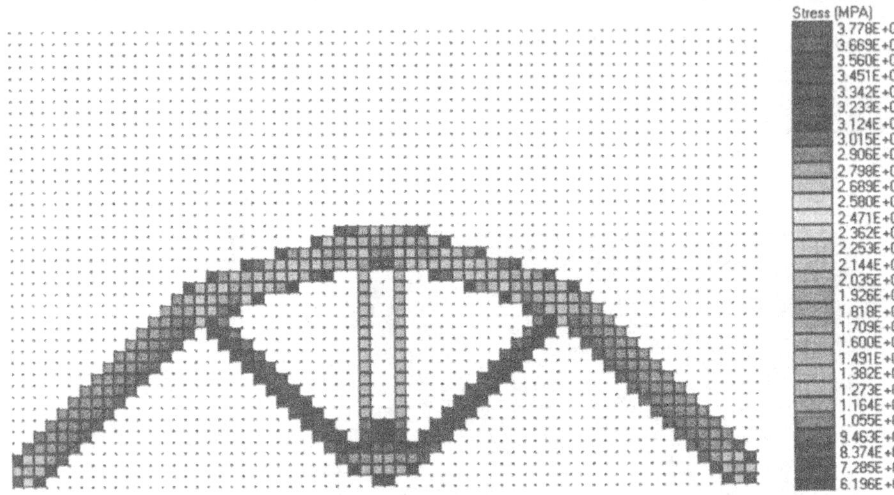

Fig. 9.35. Michell structure at Rejection Ratio = 15%.

Congratulations!

You have just optimized your first structure with the ESO method and Evolve97!

9.4.6 Viewing the Evolution History

Now that you have an optimum structure you probably want to see how the stresses and volume have change during the evolution. To do this follow these steps:

- Press the Plot Evolution History button.
- The dialogue box of Figure 9.13 should appear.
- Select the Stress vs Rejection Ratio radio button.
- Then press the OK button.
- The chart of Figure 9.36 should appear.
- And if you want to see the stress and volume ratios vs *RR*, press the Plot Evolution History button again.

- Select the Vol & Stress Ratio vs Rejection Ration radio button.
- Press OK.
- The chart of Figure 9.37 should appear.

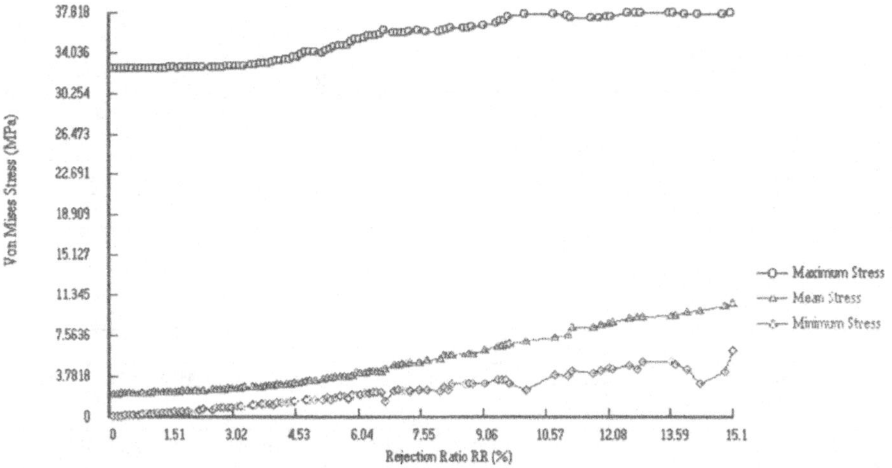

Fig. 9.36. Chart of maximum, mean and minimum stress evolution history.

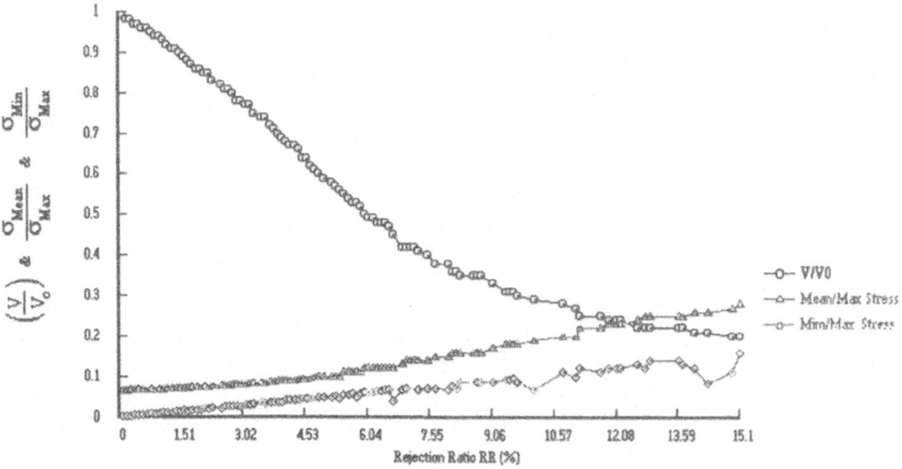

Fig. 9.37. Chart of stress and volume ratios evolution history.

9.5 Evolve97 Data Storage Files

This section contains a table detailing the files that are generated and used by Evolve97. These extensions are appended to the name of the structure.

File extension	Description
~ca	Back up of 'nca' file
~do	Back up of 'ld0' file
~de	Back up of 'ide' file
~dg	Back up of 'edg' file
~lt	Back up of 'plt' file
~pr	Back up of 'ppr' file
~re	Back up of 'fre' file
~rr	Back up of 'err' file
edg	Contains the number of plate edges connected to other plate elements
elg	Contains the structure evolution log, in ASCII format. It is the data plotted by the 'Plot Evolution History of Structure' button.
err	Contains the Rejection Ration value when plate elements are removed.
eso	Contains the Evolutionary Structural Optimization parameters
est	Contains the stresses in each of the plate elements
evl	Contains general information on the number of plates, nodes, freedoms and loads on the structure.
fre	Contain the freedom conditions for all the nodes in the structure
ide	Contains the structure's identity matrix. Used by the FE solved
ld0	Contains the loads applied to all the nodes in the structure
mo2	Original plate connectivity file. It is a record of the elements present before the evolution process was started
nca	Contains information on the number of the plates attached to each node
nds	Contains the node screen coordinates.
nod	Contains the coordinates of every node in the structure
pdp	Contains the displacement of each node due to the applied loads
plc	Contains the coordinates of every plate's centre in screen coordinates
plt	Plate connectivity file
ppr	Contains the material properties of each plate element

9. 6 Error Messages

This section is intended to give you some assistance in the event that Evolve97 experiences some errors. There is only one Evolve97 error message that can appear, although in rare circumstances Evolve97 can raise some Windows 95 errors.

ERROR_FESolver_Zero

This error is raised by the finite element solver when one of the terms in the diagonal of the global stiffness matrix has a zero value and therefore the equations cannot be solved.

The most common reason for this error is not including the material properties in the Specify Design and Non-Design dialogue boxes, shown in Figures 9.19 and 9.20.

It may also arise at the end of the evolution process when most of the plate elements have been removed. In rare cases plate elements may be left unconnected to the rest of the structure. This will also cause the global stiffness matrix to have zero terms in the diagonal.

EOutOfResources

This is a Windows error. It occurs when Windows cannot fill an allocation request for a Windows resource, in this case drawing to the screen when the structure is evolving. To prevent this from happening, make sure the Zoom Factor is equal to 0.6 or 0.7 when you evolve a structure. You can use smaller factors. In doing so this error should not occur.

Although Evolve97 is terminated by Windows when this error occurs, you can re-start Evolve97 again without having lost any data. You can even start the optimization problem from where Evolve97 left off when it was terminated by Windows. Just draw the structure and you will realize this.

EOutOfMenory or **ERROR_NOT_ENOUGH_MEMORY**

These are Windows errors. They indicate that your computer does not have enough free memory available to run Evolve97. Make certain that you have at least 50 MB free Hard Disk space and at least 16 MB of RAM.

If you have any difficulties in using Evolve97 or have queries about any part of this book, please contact us by E-mail at the following addresses:

ozz@aero.ae.su.oz.au (O.M. Querin)
mikexie@vut.edu.au (Y.M. Xie)
grant@aero.ae.su.oz.au (G.P. Steven)

Author Index

Subject Index

Made in the USA
Monee, IL
13 September 2026

62976974R00116